SO-BYH-195

MOLECULAR
BIOLOGY
INTELLIGENCE
UNIT

Signal Transduction by Integrins

Paola Defilippi
Guido Tarone
Department of Genetics, Biology and Medical Chemistry
University of Torino
Torino, Italy

Angela Gismondi
Angela Santoni
Department of Experimental Medicine and Pathology
University of Rome
Rome, Italy

CHAPMAN & HALL
I⊕P An International Thomson Publishing Company

New York • Albany • Bonn • Boston • Cincinnati • Detroit • London • Madrid • Melbourne •
Mexico City • Pacific Grove • Paris • San Francisco • Singapore • Tokyo • Toronto • Washington

AUSTIN, TEXAS
U.S.A.

MOLECULAR BIOLOGY INTELLIGENCE UNIT
SIGNAL TRANSDUCTION BY INTEGRINS

LANDES BIOSCIENCE
Austin, Texas, U.S.A.

U.S. and Canada Copyright © 1997 Landes Bioscience and Chapman & Hall

Please address all inquiries to the Publishers:
Landes Bioscience, 810 South Church Street, Georgetown, Texas, U.S.A. 78626
Phone: 512/ 863 7762; FAX: 512/ 863 0081

North American distributor:

Chapman & Hall, 115 Fifth Avenue, New York, New York, U.S.A. 10003

CHAPMAN & HALL

U.S. and Canada ISBN: 0-412-13301-6

Library of Congress Cataloging-in-Publication Data

Signal transduction by integrins / Paola Defilippi
     p. cm. — (Molecular biology intelligence unit)
   Includes bibliographical references and index.
   ISBN 1-57059-473-2 (alk. paper)
   1. Integrins — Physiological effect. 2. Cellular signal
transduction. I. Defilippi, Paola, 1957–. II. Series.
    [DNLM: 1. Integrins — physiology. 2. Signal Transduction –
– phsiology. QW 570 S578 1997]
QP552.I55S54 1997
572'.68—dc21
DNLM/DLC
for Library of Congress   97-24582
                                                     CIP

PUBLISHER'S NOTE

Landes Bioscience produces books in six Intelligence Unit series: *Medical, Molecular Biology, Neuroscience, Tissue Engineering, Biotechnology* and *Environmental.* The authors of our books are acknowledged leaders in their fields. Topics are unique; almost without exception, no similar books exist on these topics.

Our goal is to publish books in important and rapidly changing areas of bioscience for sophisticated researchers and clinicians. To achieve this goal, we have accelerated our publishing program to conform to the fast pace at which information grows in bioscience. Most of our books are published within 90 to 120 days of receipt of the manuscript. We would like to thank our readers for their continuing interest and welcome any comments or suggestions they may have for future books.

Shyamali Ghosh
Publications Director
Landes Bioscience

CONTENTS

PREFACE

Matrix biology initially focused on matrix protein structure and assembly to investigate how cells are held together to form tissues. Since matrix composition and architecture varies considerably in different tissues, such as muscle, cartilage, bone and skin, it was hypothesized that matrix may also affect cell behavior. This concept was supported by a series of experimental evidence but was enormously boosted by two major findings: the discovery of integrins as cell surface receptors for matrix proteins and the demonstration of their signaling capacity. The discovery of integrins revealed that they not only function in cell-matrix interactions but also in cell-cell adhesion and are involved in a wide variety of biological processes. Moreover, the finding that integrins generate intracellular signals opened a new avenue of research aimed to investigate the molecular basis by which cellular interactions can affect cell behavior. Recently this field literally exploded with almost 200 papers published on integrin signaling in the last year and it is still rapidly evolving. The picture emerging from these studies indicate that integrin signaling intersects with signaling from a number of other surface receptors explaining how adhesive interactions can represent consensus signals for several cellular functions including proliferation, differentiation and migration.

In this book we summarize the available literature on integrin signaling and make an effort to combine information that is frequently fragmentary or contradictory. We apologize for the work that has not been quoted in the references; these omissions are not intentional but due to the vast amount of material published. We discussed in the first chapter general aspects of integrin structure and function and, in the following chapters, the distinct pathways involved in the transmission of the signal from the membrane to the nucleus. The last chapter is devoted to the complex crosstalk between integrins and cell surface receptors in the hematopoietic system.

CHAPTER 1

Integrin Structure and Function

Integrins are a class of membrane-spanning glycoproteins that functionally link the cytoskeleton to the extracellular environment (Hynes 1987). These molecules, consisting of a heterodimer of noncovalently linked α and β subunits, bind to a variety of extracellular ligands including extracellular matrix proteins, complement components and cell surface adhesive molecules. Following their binding to their extracellular ligands, integrins transduce signals that lead to linkage and organization of cytoskeletal proteins at the cytoplasmic surface. Most integrins connect to bundles of actin filaments via bridging proteins such as vinculin, talin, tensin and paxillin, the major exception being the β4 integrin that connects to intermediate filaments. The interaction with the actin cytoskeleton leads to redistribution of integrins at the cell surface and to their clustering in specific structures known as focal adhesions. Because of this linkage across the plasma membrane, the cytoskeleton can anchor to the extracellular environment and control the tensional forces necessary for cell movement, cell-cell interactions and organization of cells in tissues during both embryonic development and tissue repair. Several genetic defects demonstrate the importance of integrins in these processes. Mouse embryos show lethal defect at various stages of development when expression of β1, β4, α3, α4, α5, α6, α8 or α9 integrins is abrogated by gene targeting (Fässler et al 1996). In *Drosophila* mutants lacking expression of βps, myotendinous junctions fall apart at the first attempt of muscle contraction (MacKrel et al 1988). In humans, lack of expression of the fibrinogen receptor αIIbβ3 in platelets leads to serious defects in blood clotting (Perutelli and Mori 1992) and absence of the β2 integrins causes defect in leukocyte extravasation during inflammatory response leading to recurrent bacterial infections (Wardlaw et al 1990). In addition to organizing the cytoskeleton and transducing mechanical forces across the membrane, integrins also send signals inside the cell that control cellular metabolism, gene expression and proliferation. The ability of integrins to generate intracellular signals that control assembly of the cytoskeleton and cell behavior will be addressed here. Before discussing these aspects, however, we will analyze the receptor family and the molecular basis of integrin function.

THE DISCOVERY OF THE INTEGRIN FAMILY

One of the most intriguing discoveries after isolation of the fibronectin receptor (Pytela et al 1985a) was the unpredicted finding that this molecule has common structural features with the fibrinogen receptor in platelets, with the so-called position-specific antigens in *Drosophila* involved in determining cell fate during development of imaginal discs, as well as

Signal Transduction by Integrins, by Paola Defilippi, Angela Gismondi, Angela Santoni and Guido Tarone. © 1997 Landes Bioscience.

with leukocyte antigens involved in the inflammatory and immune responses (Leptin 1986). Determination of the primary sequence proved that these receptors belong to a family of structurally related proteins with common evolutionary origin termed integrins (Hynes 1987). The integrin family in mammals now includes eight different β and sixteen different α subunits that can associate in heterodimers forming a variety of different complexes. Twenty-two α/β integrin complexes occur in nature (Table 1.1).

THE β1 GROUP

Ten different α subunits can associate to β1 forming ten different receptors. Most of these receptors bind to matrix ligands, such as fibronectin, laminin and collagen, but the α4β1 and α5β1 complexes can also bind to VCAM-1 and L1, respectively (Ruppert et al 1995), two adhesive cell surface receptors of the Ig superfamily. The β1 integrins are ubiquitous molecules expressed on a variety of different cell types, the major exception being circulating red blood cells. The β1 integrin is absolutely required for embryonic development as shown by gene knockout experiments (Fässler and Meyer 1995; Stephens et al 1995). Embryonic lethal defects are also observed in mice lacking expression of the associated α4, α5, αV subunits. On the other hand, the α1, α3, α6, α7, α8 and α9 subunits are not crucial for embryonic development, but their absence causes different genetic defects in newborn or adult mice (Fässler et al 1996).

THE β2 GROUP

The β2 integrins are expressed exclusively in leukocytes. Four different α subunits (van der Vieren 1995; Danilenko et al 1995) can associate to β2 forming receptors that recognize a variety of ligands including complement component C3bi, fibrinogen, coagulation factor X and the adhesive cell surface receptors, ICAM-1, ICAM-2 and ICAM-3, as well as lipopolysaccharide, heparin and haptoglobin (Diamond et al 1995; Ingalls and Golenbock 1995; El Ghmati et al 1996). These receptors play a crucial role in inflam-

mation and immunity by controlling the adhesive behavior of neutrophils, monocytes/macrophages and lymphocytes. The vital importance of β2 integrins is highlighted by the LAD (*L*eukocyte *A*dhesion *D*eficiency) syndrome in human and cows (Harlan 1993; Olchowy et al 1994). This is an autosomal recessive disease characterized by genetic mutations in the β2 gene leading to lack of expression of β2 integrins. Individuals affected by LAD syndrome are subjected to recurrent infections due to the inability of leukocytes to extravasate and reach the site of inflammatory reaction.

THE β7 GROUP

Two α subunits can associate with β7 integrin. The αEβ7 and α4β7 integrins are specifically expressed on a subset of lymphocytes and macrophages (Tiisala et al 1995) and mediate their homing to the lymphoid organs of the gut and adhesion to the epithelial cells (Wagner et al 1996). αEβ7 binds to E-cadherin, while α4β7 binds to fibronectin as well as to the adhesive cell surface receptors, VCAM-1 and MADCAM (Cepek et al 1994).

THE αV GROUP

This group is identified by the α subunit, rather than by the β as in previous cases, since αV can associate with several different β subunits including β1, β3, β5, β6 and β8. The prototype of this group is the αVβ3 complex originally identified as the vitronectin receptor (Pytela et al 1985b). αVβ3 can bind to several different matrix ligands (see Table 1.1) and to the PECAM-1 cell surface receptor (Piali et al 1995). αVβ3 is expressed in most cultured cell lines, but osteoclasts express the highest levels in vivo. In these cells αVβ3 has an important role in the formation of the osteoclast-bone matrix contact during bone resorption (Crippes et al 1996).

OTHER INTEGRINS

Three additional integrin complexes, the αIIbβ3, the α6β4 and the leukocyte response integrin (LRI), cannot be classified in the previous groups.

Table 1.1. Integrin family of adhesion receptors

Dimers	Alternative Nomenclature	Ligands	Distribution
α1β1 (*)	VLA1, CDw49a/CD29	Laminins (E1), [1] Collagen I, IV	Small vessel endothelium, smooth muscle, fibroblasts
α2β1 (*)	VLA2, CDw49b/CD29	Laminins (E8), [1] Collagen	Epithelia, platelets, fibroblasts
α3β1	VLA3, CDw49c/CD29	Laminins (E8), Nidogen, Fibronectin (Arg-Gly-Asp), [2] Collagen, Invasin	Smooth muscle, connective tissue, keratinocytes, kidney
α4β1	VLA4, CDw49d/CD29	Fibronectin (Leu-Asp-Val), [2] VCAM-1	Leukocytes, developing skeletal muscle
α5β1	VLA5, CDw49e/CD29	Fibronectin (Arg-Gly-Asp), L1, Invasin	Widespread
α6β1	VLA6, CDw49f/CD29	Laminins (E8), Fertilin, Invasin	Epithelia, vessel wall, nerves, oocytes
α7β1	CDw49/CD29	Laminins (E8)	Skeletal muscle, vessel wall
α8β1	CDw49/CD29	Tenascin	Nervous tissue, epithelia, muscle
α9β1	CDw49/CD29	Tenascin	Epithelia and muscle
αLβ2 (*)	LFA-1, CD11a/CD18	ICAM-1, ICAM-2, ICAM-3	Lymphocytes, myeloid cells
αMβ2 (*)	Mac-1, CD11b/CD18	C3bi, Fibrinogen, Heparin, Haptoglobin, Hemagglutinin (Bord. Pertussis)	Myeloid cells, lymphocyte subsets
αXβ2 (*)	gp150/95, CD11c/CD18	C3bi, Fibrinogen	Myeloid cells, lymphocyte subsets
αDβ2 (*)	CD11d/CD18	ICAM-3, ICAM-1	Tissue macrophages
αEβ7	αielβ7, αHβ7, αm290β7	E-Cadherin	Lymphocytes, Macrophages
α4β7	CDw49d/	Fibronectin (Leu-Asp-Val), VCAM-1, MADCAM	Lymphocytes

(continued on next page)

Table 1.1. Integrin family of adhesion receptors (cont.)

Dimers	Alternative Nomenclature	Ligands	Distribution
αVβ1	CD51/CD29	Fibronectin, Vitronectin (Arg-Gly-Asp)	Fibroblasts, neuroblastoma, 293 kidney cells in vitro
αVβ3	CD51/CD61	Vitronectin (Arg-Gly-Asp), Fibrinogen, von Willebrand, Fibronectin (Arg-Gly-Asp), Denatured collagen and laminin, Osteopontin, Tenascin, Thrombospondin, PECAM-1	Osteoclasts, endothelium, connective tissue
αVβ5	CD51/	Vitronectin	Epithelia
αVβ6	CD51/	Fibronectin, Tenascin	Epithelia
αVβ8	CD51/	?	?
αIIbβ3	CD41/CD61, GPIIb/IIIa	Fibrinogen, von Willebrand, Vitronectin, Fibronectin (Arg-Gly-Asp)	Platelets
α6β4	CDw46f/	Laminins	Epithelia, Schwann cells, thymocytes
α(?)β(?)	LRI (leukocyte response integrin)	Fibrinogen, fibronectin, von Willebrand, vitronectin, collagen, entactin, (Arg-Gly-Asp)	Monocytes, polymorpho nuclear cells and lymphocytes

(*) the α subunits of these complexes contain the I domain (see Fig. 1.1).
(1) Arg-Gly-Asp and Leu-Asp-Val correspond to the two adhesive sites of the fibronectin molecule.
(2) E1 and E8 are two different regions of the laminin 1 molecule corresponding to the center of the cross and to the end of the long arm respectively (Timpl and Brown 1994).

αIIbβ3 is the fibrinogen receptor in platelets. This molecule was known as the GPIIb/IIIa complex well before the discovery of the integrin family and its function in platelet thrombus formation during blood clotting has been extensively investigated. Platelets are the only cells expressing this integrin complex in vivo. A genetic disease, Glanzmann thrombastenia, is characterized by mutations in the αIIb or β3 genes which cause lack of expression or misfunction of this receptor with consequent severe blood clotting disorders (Perutelli and Mori 1992).

The α6β4 is a receptor for laminins expressed predominantly in epithelial cells. Lower level of expression occurs also in a subset of endothelial cells, in immature thymocytes and in Schwann cells. In stratified epithelia α6β4 is expressed only on the ventral surface of the basal layer in opposition to the basal laminae and is concentrated in the hemidesmosomes. These are junctional structures linking keratin intermediate filaments to the extracellular matrix. β4 is the only integrin known to connect intermediate filaments to the plasma membrane while all other integrins are involved in mediating the actin microfilament-membrane linkage. β4 also differs from other integrins in the cytoplasmic domain that in this protein is 1000 amino acid residues long. This molecule is absolutely required for the stable adhesion of the epithelia to the basal lamina and the underlying derma as demonstrated by knockout experiments and by the presence of mutation of the β4 gene in patients affected by a severe form of epidermolysis bullosa, a syndrome causing diffuse skin blistering (Vidal et al 1995; van der Neut et al 1996; Dowling et al 1996).

The Leukocyte Response Integrin (LRI) is expressed on polymorphonuclear cells, monocytes and lymphocytes, and it recognizes fibrinogen, fibronectin, von Willebrand factor, vitronectin, collagen and synthetic peptides containing the Arg-Gly-Asp (Carreno et al 1993). Engagement of LRI on polymorphonuclear cells leads to both increased phagocytosis via Fc receptors and to adhesion and chemotaxis to certain extracellular matrix proteins. The molecular identity of this integrin remain to be determined since these molecules have not been cloned: the β subunit of LRI shares an antigenic epitope(s) with β3, but is not identical to β3 and the α subunit is unknown.

MANY INTEGRINS OCCUR IN SEVERAL SPLICING FORMS

The complexity of the integrin family is further increased by the presence of different splice variants of several subunits. The first splice variants identified were the cytoplasmic domain isoforms β3B (van Kuppevelt et al 1989) and β1B (Altruda et al 1990). The β1B isoform is generated by usage of an alternative polyadenylation site present in a 3' intron (Altruda et al 1990; Baudoin et al 1996) which causes lack of splicing and retention of this intron in mature mRNA. The β3B isoform is likely to be generated by similar mechanisms, although the genomic structure of this region has not been determined. Two more variants of the β1 have been identified, β1C and β1D, which are generated by alternative exon usage (Languino and Ruoslahti 1992; Zhidkova et al 1995; van der Flier et al 1995; Belkin et al 1996). In all four β1 isoforms, splicing occurs at the end of exon 6 that codes for the transmembrane and part of the cytoplasmic domain. Thus, the cytoplasmic domain of the four isoforms consists of a shared membrane proximal region coded for by exon 6 (common region) and a COOH terminal region unique to each isoform coded for by exons B, C or D (variable region) (see Fig. 1.3). In addition to β3B, a second β3 variant lacking the transmembrane and cytoplasmic domain has been described (Djaffar et al 1994).

The β4 cytoplasmic domain is found in different forms generated by alternative splicing. Three extra sequences of 53 (Hogervorst et al 1990), 70 (Tamura et al 1990) and 7 (Clarke et al 1994) amino acid residues respectively have been identified that potentially generate a variety of different isoforms.

Alternative splicing also occurs in α subunits, and isoforms of α3 (Tamura et al 1991), α6 (Cooper et al 1991; Hogervorst et al 1991) and α7(Collo et al 1993; Ziober et al 1993) have been described. In α6 and α7 subunits, isoforms differing in the cytoplasmic region as well as in the extracellular portion close to the divalent cation binding motifs have been identified (Delwel et al 1995; Ziober et al 1993).

With the exception mentioned above, all known variants differ in the cytoplasmic domain sequence. This is a rather peculiar feature since the cytoplasmic domain of integrins are the most conserved portion of the molecule during evolution. In the case of β1, all 41 residues are identical in frogs and humans (Sastry and Horwitz 1993). Similarly α5 cytoplasmic domain residues are identical in mice and humans (Sastry and Horwitz 1993). Thus, the folding and molecular interactions of the cytoplasmic domain are so strictly regulated that even single residue mutations were selected against during evolution. In spite of this fact, alternative splicing forms have evolved leading to the synthesis of proteins with divergent sequences. This suggests that the alternative forms should have distinct and specific functions. Functional analysis of the four β1 integrin isoforms shows that this is indeed the case. The β1B has the unexpected properties of an anti-adhesive receptor: it does not activate signaling through p125Fak nor does it organize focal adhesions (Balzac et al 1993). If coexpressed with other integrins, such as β1A or β3, it has a dominant negative action and inhibits cell spreading, migration, assembly of fibronectin matrix and formation of focal adhesions (Balzac et al 1994; Retta et al submitted). β1B is expressed in keratinocytes and hepatocytes at low levels and simultaneously with the β1A isoform (Balzac et al 1993). Its function in these tissues is presently unknown. The dominant negative action suggests that β1B may be important in regulating de-adhesion during cell migration and differentiation.

β1C also has unexpected function: its expression in cultured fibroblasts arrests cell cycle progression at the G1/S boundary (Meredith et al 1995; Fornaro et al 1995). This molecule is expressed at low levels in endothelial cells induced to quiescence by TNF treatment, suggesting a possible physiological role (Fornaro et al 1995).

β1D is specifically expressed in skeletal muscles and heart (Zhidkova et al 1995; van der Flier et al 1995; Belkin et al 1996) and it represents the only β1 isoform in these tissues. β1D has a transdominant effect on β1A. In cells expressing both isoforms β1D localizes at focal adhesions displacing β1A from these structures. β1D has a higher capacity to bind matrix ligands and cytoskeletal components of the focal adhesions such as talin. These properties allow β1D to form more stable matrix-cytoskeleton contacts that are necessary to support the strong mechanical tension during muscle contraction (Belkin et al submitted).

Distinct patterns of expression as well as different functional properties have been described for the splicing variants of other integrin subunits. α6B is expressed in early development and has a much wider distribution as compared with α6A (Cooper et al 1991; Hogervorst et al 1993a; Thorsteinsdottir et al 1995). α6A is more efficient in triggering tyrosine phosphorylation of paxillin and other proteins (Shaw et al 1995b), but contradictory results have been reported concerning the ability of the two variants to support adhesion to laminin isoforms (Delwel et al 1993; Shaw and Mercurio 1994). In addition α6A and α6B have a differential distribution in the adhesive structures in the ventral plasma membrane of adherent embryo fibroblasts (Cottellino et al 1995). α7B, as α6B, is expressed early during development, while α7A appears later and is selectively expressed in striated muscles (Collo et al 1993; Velling et al 1996). Moreover, α7A, α7B and α7C isoforms seem to have a distinct sub-cellular distribution in skeletal myofibers (Martin et al 1996).

SEVERAL DIFFERENT RECEPTORS CAN BIND TO THE SAME LIGAND

It has been shown that often several different integrins can bind to the same matrix ligand. Eight different fibronectin recep-

Table 1.2. Multiple integrins bind to the same extracellular matrix ligands

	α1β1	α2β1	α3β1	α4β1	α5β1	α6β1	α7β1	αVβ1	αVβ3	αVβ6	αVβ5	α6β4	α4β7	αIIbβ3
Fn			•	•	•			•	•	•			•	•
Lm	•	•	•			•	•		•			•		
Vn								•	•		•			•
Coll	•	•	•						•					

tors have been described, namely α3β1, α4β1, α5β1, αVβ1, αVβ3, αVβ6, αIIbβ3 and α4β7. Similarly, α1β1, α2β1, α3β1, α6β1, α7β1, αVβ3 and α6β4 all bind to laminin (Table 1.2). The number of possible fibronectin and laminin receptors is further increased by the presence of alternative splicing forms of both α and β subunits. This multiplicity of receptors indicates that evolution has provided a complex compensatory system. Gene knockout experiments indicate that this is the case for several integrins. One of the most clear examples is the α5β1 fibronectin receptor. Cells lacking α5β1 still assemble a fibronectin matrix in vitro utilizing other β1 (Yang et al 1993) or β3 (Wennenberg et al 1996) integrins. In fibroblasts lacking the common β1 subunit, adhesion to fibronectin-coated dishes occurs via the αVβ3 complex (Wennenberg et al 1996). This phenomenon is likely to occur also in vivo; in fact, mice lacking α5 subunit die at a later stage of development as compared with mice lacking fibronectin, indicating that the absence of the ligand causes a more severe phenotype than the absence of the receptor. This finding suggests that α5β1 is substituted by other fibronectin receptors in vivo allowing a more prolonged survival (Yang et al 1993). Evolution has, thus, favored the development of a redundant system to increase the chances of overcoming specific lack-of-function mutations.

This aspect is further emphasized by the ability of given integrin complexes to bind multiple ligands. The αVβ3 complex is the most explicit example of this property. Ten

different ligands, including vitronectin, fibronectin, fibrinogen and several others (see Table 1.1) can bind to αVβ3. The ability to bind these molecules can be affected by the membrane environment as indicated by the fact that phospholipids affect the binding specificity of αVβ3 (Conforti et al 1990). Thus receptors such as αVβ3 are particularly suitable for rescue function.

In several cases, distinct receptors are able to elicit different cellular responses. This has been clearly demonstrated in several cases indicating that not all receptors for a given matrix protein are suitable for rescue function. The α4β1 and α5β1 receptors recognize two distinct sites in fibronectin and trigger specific responses. While engagement of α5β1 with specific fibronectin fragments induces collagenase synthesis, α4β1 does not and actually inhibits this response (Huhtala et al 1995) (see chapter 7). αVβ3 and αVβ5 bind to vitronectin but have a different role in cell migration (Klemke et al 1994; Liaw et al 1995). αVβ1 and α5β1 can both bind to fibronectin, but differ in their capacity to promote fibronectin matrix assembly (Zhang et al 1993). In the case of laminin, α6β1 and α6β4 are obviously different as the former one organizes the actin filaments, while the latter is responsible for membrane-cytokeratin filaments interaction. The two laminin receptors α1β1 and α6β1 differ in their ability to activate the MAPK pathway via Shc (Wary et al 1996). Thus, the presence of a large number of different receptors for a given matrix molecule has a dual importance: firstly to allow the differentiation of the cellular response and secondly, to provide a compensatory system to rescue lethal defects.

A given cell type usually expresses multiple receptors for a single ligand, raising the question of how these receptors are regulated in their function at the cell surfaces. Recent data indicate that different integrin heterodimers affect each other and a hierarchy in their function is frequently established. We have previously mentioned that $\alpha V\beta 3$, that normally acts as a vitronectin receptor, can substitute $\alpha 5\beta 1$ in fibronectin matrix assembly and cell adhesion when the latter one is missing (Wennenberg et al 1996; Retta et al submitted). However, when both receptors are present, $\alpha 5\beta 1$ acts as dominant fibronectin receptor, limiting or abrogating $\alpha V\beta 3$ binding to fibronectin (Retta et al submitted). Moreover, the $\beta 1B$ and $\beta 1D$ isoforms affect $\beta 1A$ function since they act as a dominant negative (Balzac et al 1994) and dominant positive receptors (Belkin et al submitted), respectively.

INTEGRIN EXPRESSION AT THE CELL SURFACE IS REGULATED BY EXTRACELLULAR STIMULI

Integrin expression at the cell surface can be modified by several stimuli. The first example reported concerns the ability of TGFβ to induce upregulation of β1 integrins. This parallels the stimulation of matrix protein synthesis by TGFβ and is related to the ability of this cytokine to restore anchorage-dependent cell growth (Ignotz and Massague 1987). Several other cytokines and differentiation factors are also able to modify integrin expression in different cell types and thus to regulate the adhesive behavior during both physiological and pathological processes. The abundant literature on this aspect cannot be exhaustively reviewed here. A few examples of molecules known to modify integrin expression at the cell surface include cytokines such as GM-CSF, TNFα, IL1-β, IFNγ, MCP-1 and IL-4 (Socinski et al 1988; Defilippi et al 1991; Santala and Heino 1991; Jiang et al 1992; Kitazawa et al 1995); growth and differen-

tiation factors, such as NGF, PDGF and retinoic acid (Rossino et al 1990, 1991; Xu and Clark 1996) and oncogenes (Plantefaber and Hynes 1989; Inghirami et al 1990).

MOLECULAR STRUCTURE OF α AND β SUBUNITS

Integrins are heterodimers of noncovalently associated α and β subunits. Both α and β subunits are membrane-spanning glycoproteins with a large extracellular region and a short cytoplasmic domain. Several different α and β subunits have been identified and sequenced and characteristic primary structure features have been identified.

The α subunits are approximately 1050 amino acids long and most of the sequence is exposed at the extracellular space, the cytoplasmic domain consisting only of 15-50 amino acid residues. α subunits can be classified in two groups according to their structural features (Fig. 1.1). The subunits of the first group are cleaved in heavy and light chains that are held together by a disulfide bridge. The heavy chain (approximately 120,000 dalton) is entirely extracellular, while the light chain (approximately 25,000 dalton) spans the plasma membrane. An exception is represented by the α4 subunit in which the cleavage site occurs toward the middle of the molecule and generates two 80,000 and 70,000 dalton chains. The second group is comprised by single chain molecules containing an extra sequence of 180-200 amino acid residues homologous to the A domain of von Willebrand factor called the I or A domain (Hemler 1990). This region is important in ligand binding as discussed below. All α subunits also contain multiple repeats of a sequence homologous to the EF hand motif (Kawasaki and Kretsinger 1995) with the ability to bind cations. Four of these repeats are present in the α subunits of the first group and three in those of the second one. Cations are important in regulating integrin-ligand binding (Gailit and Ruoslathi 1988). Cation binding sites are present also in the I domain and their role in ligand binding has

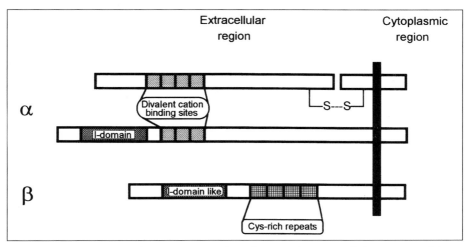

Fig. 1.1. The structure of the integrin α and β subunits. The extracellular portion of the α subunits is characterized by the presence of a repeated motif homologous to the EF hand structure capable to bind divalent cations. A group of α subunits is characterized by the presence of the I domain, an inserted sequence of approximately 200 amino acid residues, important in ligand binding. This group consists of a single polypeptide chain and is not processed into heavy (extracellular) and light (transmembrane) chains linked by a disulphide bridge. The β subunit is characterized by four cysteine-rich repeats and by an I-domain like sequence involved in ligand binding. The I-domains of α and β subunits can bind divalent cations.

been documented (Michishita et al 1993).

The β subunits are approximately 780 amino acid residues long and contain 56 cysteine residues whose positions are highly conserved among different β subunits as well as between *Drosophila* and human molecules (Hemler 1990). A large portion of the molecule, in general 700-730 residues, is exposed at the extracellular surface and only 40-50 residues are located in the cytoplasmic domain (Fig. 1.1). The β4 subunit is a notable exception as it consists of 1752 residues, 1000 of which contribute to the cytoplasmic domain (Suzuki and Naitoh 1990; Hogervorst et al 1990; Tamura et al 1990). Elucidation of the disulfide bonding pattern in the β3 subunit (Calvete et al 1991) showed the presence of two bonds, linking residues 5-435 and 406-655, respectively, and forming two long-range loops in the extracellular domain. The remaining cysteine residues are grouped in a 4-fold repeated cysteine-rich segment located between residues 406

and 655. These residues form extensive disulfide bridging and the precise pattern has not been unambiguously determined. Overall, the disulfide bonds lead to a highly packed three-dimensional structure that is responsible for the high resistance to protease attack (Tarone et al 1982; Giancotti et al 1985). Detailed sequence analysis has provided evidence for the presence of a region N-terminus to the cysteine-rich cluster with homologies to the I domain of the α subunit and involved in ligand binding (Smith and Cheresh 1990; D'Souza et al 1990; Loftus et al 1990; Lee et al 1995).

The α and β subunits dimerize to form a heterodimer that, by electron microscopy, appears as a globular head with two extending stalks (Nermut et al 1988). The extracellular domains of the two subunits likely contribute to the globular head region while the two stalks represent the cytoplasmic domains. The association of the two subunits is regulated by interaction of the extracellular domains, since truncated forms of α and β lacking the transmembrane and cytoplasmic domains can form heterodimers capable of ligand binding (Dana et al 1991).

INTEGRIN-LIGAND BINDING: IDENTIFICATION OF THE ARG-GLY-ASP SEQUENCE

Although extracellular matrix molecules are the most common integrin ligands, fibrinogen, complement components, cell surface Ig superfamily receptors and microorganisms (Isberg and Leong 1990; Relman et al 1990) can also act as ligands (see Table 1.1).

Definition of the fibronectin sequence involved in receptor binding pioneered studies on the mechanisms of integrin-ligand binding. The major difficulty of these studies was the low-affinity of fibronectin-receptor interaction which is characterized by an apparent Kd in the range of 0.2×10^{-6} mole/L. These studies lead to the discovery of the Arg-Gly-Asp sequence as the minimal amino acid sequence necessary for receptor binding (Pierschbacher and Ruoslahti 1984). Short peptides containing the Arg-Gly-Asp sequence can compete for the binding of the receptor to intact fibronectin allowing selective interference with adhesion processes. Alternatively, Arg-Gly-Asp peptides can be used to bind and isolate the receptor by affinity chromatography (Pytela et al 1987). It was subsequently discovered that the Arg-Gly-Asp sequence is an adhesion motif present in several matrix proteins in addition to fibronectin. Molecules which contain this motif in their primary sequence and act as integrin ligands include vitronectin, fibrinogen, collagens and mouse laminin 1. In spite of the widespread occurrence of the Arg-Gly-Asp motif, integrins maintain their ligand specificity: the α5β1 fibronectin receptor, for example, binds to the Arg-Gly-Asp sequence of fibronectin but not to that in vitronectin or collagens. Thus, the amino acid sequences surrounding the Arg-Gly-Asp motif define the specificity of the receptor-ligand inter-action (Ruoslahti and Pierschbacher 1987).

In addition to the Arg-Gly-Asp sequence, several other integrin peptide ligands have been identified. These include the Leu-Asp-Val sequence present in the IIICS region of fibronectin and recognized by the α4β1 integrin and the fibrinogen γ chain sequence from residue 400 to 411 that binds to the αIIbβ3 receptor in platelets. These sequences do not share homology with the Arg-Gly-Asp motif at the primary sequence levels, but they all seem to form a flexible turn protruding from the protein surface and to contain a critical Asp residue.

THE LIGAND BINDING POCKET IN THE INTEGRIN MOLECULE

Soluble peptides with the Arg-Gly-Asp sequence have been used in affinity labeling and crosslinking experiments to define the ligand binding site in αVβ3 and αIIbβ3 integrin molecules. Both α and β subunits are affinity labeled in these experiments, suggesting that both subunits are required to form the ligand-binding pocket (Smith and Cheresh 1990; D'Souza et al 1990). Sequence analysis of affinity-labeled peptides obtained by tryptic digestion of the intact molecule allowed identification of the first three cation binding domains in the α subunit and an NH_2 terminal region between residues 109-119 in the β subunit as the regions coming in contact with the ligand. A second region of contact with ligand in the β subunit was identified by a different experimental strategy involving the use of synthetic peptides derived from the β subunit to interfere with ligand binding (Charo et al 1991). This region encompasses residues 204-229 of β3 and is part, together with the site discussed previously (residues 109-119), of the putative I domain present in all β subunits (residues 110-290 of β3). This region of the β subunit contains a highly conserved Asp-x-Ser-x-Ser motif that is part of the cation binding site in the EF hand sequence suggesting that the β subunit also binds cations. The importance of the Asp residue in this position is shown by the presence of a naturally occurring mutation in a

Glanzmann patient. The protein with this Asp-Tyr mutation folds properly and associates with αIIb, but no longer binds the ligand (Loftus et al 1990).

Direct evidence for the role of the α subunit cation-binding domains in ligand recognition comes from experiments using recombinant fragments. Gulino et al (1992) expressed the four calcium-binding domains of the αIIb subunit in bacteria showing that this isolated fragment binds to fibrinogen and the Arg-Gly-Asp peptide. Binding to these ligands, moreover, is cation-dependent in a manner analogous to that of the intact receptor. In the integrin α subunits belonging to the second group, the I domain is responsible for the ligand binding. Isolated recombinant I domains of several integrins bind to ligands mimicking the intact receptor (Diamond et al 1993; Michishita et al 1993; Kamata and Takada 1994; Randi and Hogg 1994; Zhou et al 1994). In most cases, divalent cations are required for interaction of the I domains with ligands and these ions have an important function in determining the three-dimensional conformation of the I domain (Lee et al 1995).

DIVALENT CATIONS REGULATE LIGAND BINDING PROPERTIES

Divalent cations play a crucial role in integrin-ligand binding. Indeed, not all cations have the same effect and multiple cation binding sites exist in the dimeric integrin molecule as mentioned above. A general rule emerging from these studies is that Ca^{2+} is usually inhibitory (with the exception of the αIIbβ3 and αLβ2 integrins) while Mg^{2+} or Mn^{2+} usually promote ligand binding. Different integrins have distinct cation sensitivity, an example being the binding of the αVβ3 and αIIbβ3 complexes to fibrinogen; while Ca^{2+} inhibits the binding of αVβ3, it strongly promotes that of αIIbβ3. In addition, interaction of a specific integrin heterodimer with different ligands is differentially affected by cations. Mn^{2+} was shown to stimulate the binding of the αIIbβ3 complex to linear Arg-Gly-Asp peptides, and to

inhibit binding to intact fibrinogen (Kirchhofer et al 1990). These studies show that the extracellular domain of integrins can undergo conformational changes that regulate ligand-binding. This is also supported by studies with monoclonal antibodies to different epitopes. While a number of antibodies were found to inhibit integrin-ligand binding, several of them have the opposite function and actually stimulate ligand interaction (Keizer et al 1988; Neugebauer and Reichardt 1991; Frelinger et al 1991; Kovach et al 1992; Arroyo et al 1993; Lenter et al 1993; Mould et al 1995). These "activating" antibodies recognize and stabilize a conformation of the receptor that has a higher ligand-binding capacity.

LIGAND BINDING CAN BE REGULATED BY CYTOPLASMIC SIGNALS: INSIDE-OUT SIGNALING

In addition to cations, which act on the extracellular domain of the integrin molecule, intracellular stimuli can also affect the state of integrin activation acting at the cytoplasmic side (Fig. 1.2). The αIIbβ3 fibrinogen receptor in platelets is the best described molecule illustrating this mechanism.

αIIbβ3 integrin becomes competent for fibrinogen binding only when platelets are activated by blood clotting stimuli (Bennet and Vilaire 1979). This has important physiological implications, as an activated receptor in circulating platelets would bind fibrinogen causing aggregation and diffuse thrombus formation. The ability to alternate between an inactive and active conformation for ligand binding has now been recognized as a general property of different integrin complexes (Faull et al 1993; Altieri and Edgington 1988; Crowe et al 1994). Several stimuli can regulate this property; in the case of platelets, exposure to thrombin or ADP leads to activation of the fibrinogen receptor. Lymphocytes can regulate β2 integrin ligand-binding upon antigen receptor stimulation (Dustin and Springer 1989). The importance of signaling mechanisms is further supported by the fact that the acti-

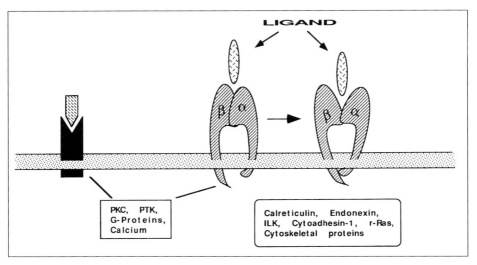

Fig. 1.2. Intracellular regulation of integrin-ligand binding capacity. In addition to divalent cations acting on the extracellular side, intracellular signals can modify integrin-ligand binding by increasing the affinity/avidity of the heterodimer for the ligand. Signals generated by membrane receptors in response to different agonists (left box) can act on integrins and modify their ligand binding capacity. The ligand binding state can be modified also by the binding of cytoplasmic molecules. The putative proteins capable of such an effect are listed in the box at right. Whether the signals generated by agonist-receptor interaction act directly or through cytoplasmic molecules such, as those listed in the box at right, is not known.

vation state of a given integrin is dependent on the cell type in which the molecule is expressed. The α5β1 fibronectin receptor is in the active state in CHO cells, but not in K562 erythroleukemia cells (O'Toole et al 1994).

This phenomenon has been referred to as inside-out signaling (Ginsberg et al 1992), to indicate that agonist-induced signals inside the cell, presumably by acting at the cytoplasmic domain, can modify the integrin-ligand binding on the outside of the cell (Fig. 1.2).

These agonists do not act directly on integrins but rather, trigger receptor-mediated signaling pathways that in turn affect integrin function. The signaling pathways involved in this phenomenon are not known. Heterotrimeric G proteins, protein kinase C, tyrosine kinases and intracellular calcium may be involved (Shattil et al 1992; van Kooyk et al 1993). Moreover, the role of protein kinase C is also suggested by the ability of PMA to induce increased adhesive

response in leukocytes and in other cell types (Show et al 1990; Pardi et al 1992). Integrin cytoplasmic domains contain consensus sites for protein kinase C phosphorylation and become phosphorylated in response to PMA stimulation. Current evidence, however, indicates that phosphorylation by protein kinase C is not the mechanism responsible for increased ligand binding in response to extracellular stimuli. In fact, the stochiometry of phosphorylation is low and mutation of the phosphorylation sites does not perturb inside-out signaling (Hibbs et al 1991; Hogervorst et al 1993b; Shaw and Mercurio 1993). Other possible cytoplasmic regulators of integrin-ligand binding are the monomeric G protein r-Ras and Rho (Laudanna et al 1996; Zhang et al 1996; see chapter 6).

Integrin cytoplasmic domain plays an important role in regulating integrin-ligand binding, presumably by imposing a conformation to the extracellular domain. Membrane proximal regions regulating integrin-ligand binding have been identified in both the α and β subunits. In the α subunit this region corresponds to the highly conserved Gly-Phe-Phe-Lys-Arg sequence and in the β subunit to the Leu-Leu-Val-Ile-Thr-Ile-His-Asp-Arg. Deletions of either one of these regions locks the integrin into the activated state (O'Toole et al 1991; O'Toole et al 1994; Hughes et al 1995; Hughes et al 1996). The function of the membrane proximal sequences is independent on signal transduction and simply reflects a change in conformation. These membrane proximal regions of the α and β subunits are, thus, involved in maintaining an inactive integrin state. Interaction of these regions with other molecules, or their deletion in the experimental conditions described above, may be responsible for a conformational change that is transferred across the membrane, switching the extracellular domain into the active state. β subunit sequences, downstream to that described above, also have an important role in regulating ligand binding. In fact, different deletions and point mutations in the COOH terminal sequences, inactivate ligand binding, indicating that COOH terminal sequences of the β cytoplasmic domain are important in regulating this function (Hibbs et al 1991; Chen et al 1992; Crowe et al 1994; O'Toole et al 1994; O'Toole et al 1995). Cytoplasmic domain splice variants of the β1 integrin also show distinct ligand-binding properties. β1 integrin cytoplasmic domain variants all share the membrane proximal region containing the Leu-Leu-Val-Ile-Thr-Ile-His-Asp-Arg motif, but differ in the distal portion toward the COOH-terminus (Altruda et al 1990; Languino et al 1992; Belkin et al 1996). We found that the β1B cytoplasmic sequence shifts the integrin complexes into the inactive state (Retta et al submitted), while the β1D sequence has the opposite effect (Belkin et al submitted).

Thus, two distinct regions of the cytoplasmic domain differentially regulate ligand binding capacity. While the membrane proximal region has a negative effect, the proximal region is a positive regulator.

It is likely that these two regions influence each other and the mechanisms that regulate the cytoplasmic domain conformation are key elements in the control of integrin-ligand binding.

A number of proteins interacting with integrin cytoplasmic domains are possible modulators of ligand binding. Cytoskeletal and signaling proteins, such as talin, α-actinin, paxillin and p125Fak, are known to bind to the β subunit cytoplasmic domain (see below) and it is possible that interaction of integrins with cytoskeletal elements stabilizes a specific cytoplasmic domain conformation that in turn modifies ligand binding (Fig. 1.2). In this regard, the binding site for p125Fak coincides almost exactly with the membrane proximal sequence Leu-Leu-Val-Ile-Thr-Ile-His-Asp-Arg known to negatively regulate ligand binding (see below and Schaller et al 1995; Hughes et al 1995). A number of other cytoplasmic proteins, not classified as cytoskeletal molecules, are also known to interact with integrin cytoplasmic domain and are likely to play a role in modifying ligand binding (Fig. 1.2). Among these, calreticulin, an intracellular Ca^{2+}-binding protein, interacts with the Gly-Phe-Phe-Lys-Arg sequence in the membrane proximal region of integrin α subunit that negatively regulates ligand binding (Rojiani et al 1991). Expression of calreticulin and its binding to integrins is necessary for cell attachment to collagen via the α2β1 receptor and for inside-out-mediated increase in ligand binding (Leung-Hagesteijn et al 1994; Coppolino et al 1995). More recently, using the "two hybrid" system, proteins that bind specifically to the cytoplasmic sequence of β3 (Shattil et al 1995), β1 (Hannigan et al 1996) and β2 (Kolanus et al 1996) integrins have been identified (Table 1.3). Endonexin is a protein that binds to the native β3 cytoplasmic domain but not to a mutated form defective in inside-out signal transduction (Shattil et al 1995), suggesting its potential role in modulation of integrin-ligand binding. ILK is a Ser/Thr kinase that associates with β1 cytoplasmic domain; its over-

Table 1.3. Molecules binding to the β1, β2, β3 integrin cytoplasmic domain

Integrin	Cytoskeletal proteins	Regulatory proteins
β1	talin, α-actinin, paxillin	p125Fak, ILK
β2	α-actinin, paxillin, filamin	p125Fak, cytoadhesin-1
β3	α-actinin, paxillin,	p125Fak, endonexin

The classification of cytoskeletal and regulatory proteins is arbitrary. See text for details.

expression reduces cell adhesion, suggesting that it negatively regulates integrin function (Hannigan et al 1996). Cytoadhesin-1, a protein with SEC7 and PH domains, binds to β2 integrins (Kolanus et al 1996) and its overexpression activates β2 integrins inducing binding to the ICAM-1 ligand. This function is supported by the isolated SEC7 domain, while the isolated PH domain inhibits β2 integrin binding suggesting a dual function of this protein.

A major question arising from these studies is whether the increased ligand binding induced by the events discussed above is due to increased affinity or avidity for the ligand. In some cases activating antibodies have been shown to increase the affinity state in a system where purified integrin heterodimers are allowed to bind monovalent ligands (Arroyo et al 1993). In most cases integrin activation was measured by cell adhesion assays or by binding of epitope-specific monoclonal antibodies without discriminating between affinity or avidity changes. Van Kooyk and co-workers (1991 and 1994) showed that both events can contribute to the increased ligand binding during activation of the αLβ2 integrin (LFA-1) in lymphocytes. These studies indicate that extracellular calcium induces aggregation of LFA-1 on the cell membrane increasing the avidity state, while other stimuli, such as PMA or triggering of the CD3/T cell receptor complex, increase affinity binding. Both events are required to induce increased LFA-1-mediated lymphocyte adhesion. These data may explain the role of both extracellular and intracellular stimuli in regulating integrin-ligand binding.

INTERACTION OF INTEGRINS WITH CYTOSKELETAL AND SIGNALING MOLECULES

Interaction of integrins with cytoskeletal molecules was originally inferred from immunolocalization studies. In cultured cells, integrins, in fact, specifically localize to focal adhesions (Damsky et al 1985; Chen et al 1986; Giancotti et al 1986) which are sites of close apposition with the growth substratum where actin filament ends are anchored to the plasma membrane (Burridge et al 1988). At these sites, integrins codistribute with cytoskeletal proteins such as vinculin, talin, α–actinin, tensin and paxillin which are part of the actin bundle-membrane linkage (Burridge et al 1988; Jockusch et al 1995). Transfection experiments, with chimeric membrane proteins bearing the extracellular domain of the IL-2 receptor and the cytoplasmic domain of integrins, demonstrated that the β, but not α, subunit cytoplasmic domain localizes at preformed focal adhesions (LaFlamme et al 1992), implying that the β cytoplasmic sequence contains all necessary information to recognize components of the focal adhesions. These findings are consistent with previous data reporting direct binding of talin with integrin complexes (Horwitz et al 1986). In these experiments equilibrium gel filtration analysis was used to detect low-affinity interactions. By this assay, the dissociation constant was calculated as approximately 10^{-6}M. Talin-integrin interaction can be displaced by soluble synthetic peptides mimicking a sequence flanking tyrosine residue 788 of the subunit cytoplasmic domain (Tapley et al 1989)

(Fig. 1.3). More recently, talin binding to β1 integrin cytoplasmic domain has been confirmed in affinity chromatography experiments; the COOH terminal portion of the β1 sequence is critical for binding in these experimental conditions (Chen et al 1995).

α-actinin is another cytoskeletal protein that binds to β subunit cytoplasmic domains. Using synthetic peptides corresponding to the β1 cytoplasmic domain adsorbed on plastic, Otey et al (1990) showed specific binding of purified α-actinin, but not vinculin. Moreover, using a series of partially overlapping decapeptides spanning the entire β1 cytoplasmic sequence, it was possible to map the α-actinin binding site to two distinct regions (Otey et al 1993) (Fig. 1.3). Interestingly, α-actinin also binds to the cytoplasmic domain of β3 and β2 integrins (Otey et al 1990; Pavalko and LaRoche 1993) and this binding is likely to be physiologically significant since α-actinin can be coimmunoprecipitated with β2 integrins in leukocytes (Pardi et al 1992; Pavalko and LaRoche 1993). In these cells the interaction of integrins with α-actinin is not constitutive, but is regulated by the cellular activation state. Association is induced in neutrophils upon stimulation with chemotactic peptides and in T lymphocytes upon antigen receptor complex stimulation.

In addition to talin and α-actinin, the actin-binding protein filamin was shown to bind directly the β2 integrin cytoplasmic domain (Sharma et al 1995) and to coimmunoprecipitate with β2 integrins in detergent cell lysate. The filamin binding site is localized to the membrane proximal region (amino acids 724-747) of the β2 tail to a site that is partially overlapping, but distinct, from the major α-actinin binding site (amino acids 733-742) (Fig. 1.3). Filamin is involved in the organization of actin filament in meshwork structure and its binding to β2 integrin may suggest a role in the formation of membrane lamellipodia during cell spreading and locomotion.

Additional proteins capable of binding integrin β subunit cytoplasmic domain are p125Fak and paxillin. Using affinity chromatography on synthetic peptides coupled to an insoluble matrix, Schaller et al (1995) showed that p125Fak binds to the first 13 membrane proximal residues of β1, a site distinct but contiguous to that of α-actinin binding (Fig. 1.3). Coimmunoprecipitation experiments have failed to show p125Fak-integrin association in vivo, thus leaving open the question as to whether integrin-p125Fak interaction is constitutive or inducible. In addition to p125Fak, paxillin, a cytoskeletal protein specifically present in focal adhesions, also binds to the same region of the β1 cytoplasmic domain. The binding, however, is not mediated by p125Fak (Schaller et al 1995).

A different approach used to investigate integrin-cytoskeleton interaction is to express integrin molecules with specific mutations or deletions in the cytoplasmic domain and assess their ability to localize in focal adhesions at the stress fiber ends (Solowska et al 1989; Marcantonio et al 1990; Reszka et al 1992). The advantage of this system is that the function is tested in vivo rather than in vitro, as for the experiments discussed above, but the major drawback is the impossibility to identify the interacting molecules. These experiments showed that removal of few amino acid residues at the COOH terminal abolished the ability of β1 and β3 to localize at focal adhesions (Marcantonio et al 1990; Ylanne et al 1995). However, more extensive truncations removing almost the entire cytoplasmic domain, but leaving three charged residues close to the membrane, still allowed partial localization (Marcantonio et al 1990). Thus both membrane proximal and distal sequences seem to have a role in integrin-actin fiber interaction. This has been confirmed and extended by a systematic mutational analysis of almost all cytoplasmic sequence residues in β1 (Reszka et al 1992). This analysis allowed identification of three regions that are important for focal adhesion localization termed cyto1, cyto2 and cyto3 (Fig. 1.3). Cyto1 is proximal to the membrane and includes the four charged residues Asp-Arg-Arg-Glu

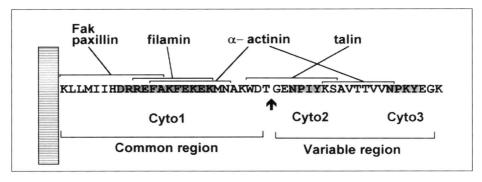

Fig. 1.3. Map of the binding sites on the β1 cytoplasmic domain sequence. The cytoplasmic domain of β1 integrin subunit consists of two region a constant regions: common to all β1 isoforms and a variable region that characterize each isoform. The arrow points to the boundary between these two regions. The sequence in the β1A isoform (one letter amino acid code) important for binding of p125Fak, paxillin, filamin, α-actinin and talin, are reported. The binding of filamin was demonstrated for β2 integrin cytoplasmic peptides and the corresponding region in β1 is indicated in the schema above. The three regions termed cyto1, cyto2 and cyto3 (shaded areas) have been identified by mutagenesis as important for localization of the protein at focal adhesions. The vertical box on the left represents the lipid bilayer of the membrane.

identified by Marcantonio et al (1990) as important for focal adhesion localization. The other two sites correspond to the two Asn-Pro-x-Tyr motifs present in the more distal region. Only mutations involving all three sites completely abolish focal adhesion localization, while single site mutations have a milder phenotype. Based on these data, a "hairpin" model has been proposed in which the COOH terminal portion interacts with the membrane proximal region by forming a β turn at the level of the Asn-Pro-x-Tyr motifs. In this structure, the cyto1 is brought in proximity of the cyto2 and 3 and the three together can represent the binding site for cytoskeletal proteins. Interestingly, β1B, a cytoplasmic domain isoform in which the cyto2 and 3 sites are substituted by an unrelated sequence, does not localize at focal adhesions and is not able to trigger p125Fak tyrosine phosphorylation (Balzac et al 1994). Similarly, truncation of few COOH terminal residues also abolishes the ability to trigger the p125Fak pathway (Guan et al 1991) indicating that this signaling event and focal adhesion localization have similar structural requirements.

As mentioned above, the cytoplasmic domain of α subunits does not have a direct role in the localization of integrins to focal adhesions (LaFlamme et al 1992). However, the α subunits can modulate the interaction of the integrin heterodimer with the cytoskeleton. In fact, while the fibronectin receptor α5β1 localizes at focal adhesions only when cells are plated on fibronectin, truncation of the α5 cytoplasmic domain allows focal adhesion localization in a ligand-independent manner (Briesewitz et al 1993; Ylanne et al 1993). Thus, the role of the α subunit cytoplasmic domain is to mask a focal adhesion localization default signal present in the β subunit so that the unligated receptor remains free in the membrane. Upon ligand binding the constraint imposed by the α subunit is released and the heterodimer can interact with focal adhesion components via the β subunit domain (LaFlamme et al 1992; Briesewitz et al 1993; Ylanne et al 1993).

INTEGRINS CAN ACTIVATE MANY DIFFERENT SIGNALING PATHWAYS

The intracellular signaling pathways triggered by integrins are directed to two major functions: organization of the cytoskeleton and regulation of cellular behavior including differentiation and growth. These two functions are not separate but are likely to be strictly interdependent (see below).

The ability of integrins to affect the above functions was inferred from a series of simple observations. Organization of the actin or intermediate filament cytoskeleton during cell adhesion clearly requires intracellular signals that trigger polymerization of cytoskeletal proteins in response to cell matrix adhesion. In addition, the ability of cell matrix interaction to control cell proliferation (Folkman and Moscona 1978) and differentiation (Menko and Boettiger 1987; Solursh et al 1984; Edgar et al 1984; Dedhar 1989) strongly implied the existence of matrix-dependent signaling events. Direct evidence for a matrix-mediated intracellular signaling was provided in the late eighties by a series of experiments showing that gene expression is altered in response to cell adhesion. Transcription of early response c-jun and c-fos genes are induced in quiescent fibroblasts upon fibronectin interaction (Dike and Farmer 1988). In monocytes, transcription of genes related to the inflammatory response is also induced by specific matrix proteins (Eierman et al 1989).

A number of reports subsequently showed that more up stream events, such as elevation of intracellular pH (Schwartz et al 1989), Ca²⁺ transients (Pardi et al 1989) and protein tyrosine phosphorylation, (Ferrel and Martin 1989) can be triggered by integrins. The list of integrin-mediated signaling events has now been expanded to comprise most of the known pathways. These include stimulation of the phosphoinositide metabolism, Ras and MAPK activation via the Grb2/Sos transducers, activation of protein kinase C, activation of tyrosine kinases and regulation of the Rho family of monomeric GTPases (see corresponding chapters).

Tyrosine phosphorylation of p125Fak has been extensively investigated and represents a characteristic feature of integrin signaling. Phosphorylation of p125Fak or of

related kinases is a common response induced by most known integrins (see chapter 2).

INTEGRIN SIGNALING REQUIRES ASSEMBLY OF A TRANSDUCTION MACHINERY AND OF ACTIN CYTOSKELETON

Although many different integrin-dependent signaling pathways have now been delineated, the molecular mechanisms by which integrins can trigger these events are still poorly defined. The cytoplasmic domain of these receptors is very short and is not endowed with enzymatic activity. Thus, interaction with transducing proteins is required to start a signaling event. The nature of these proteins, however, is still uncertain. p125Fak is a likely candidate as it binds to isolated β1 and β3 cytoplasmic domain but interaction with intact integrin molecules remains to be demonstrated (Schaller et al 1995). The ILK Ser/Thr kinase that coprecipitates with β1 integrin heterodimers and binds to the isolated β1 cytoplasmic domain (Hannigan et al 1996) may also be a good candidate to transduce signals. The β4 integrin cytoplasmic domain contains two closely spaced tyrosine residues (Immune receptor Tyrosine-based Activation Motif: ITAM sequence), similar to those found in the antigen receptor complex of lymphocytes that represent a putative anchorage site for tyrosine kinases involved in the generation of primary signaling events (Mainiero et al 1995). All integrins except β4 have a very small cytoplasmic domain that likely allows to accommodate a maximum of two cytoplasmic proteins. Considering that interaction with cytoskeletal proteins is obligatory, thus, triggering of the complex signaling pathways mentioned above requires involvement of docking/adapter proteins to allow transduction machinery assembly. p125Fak may perform this function as it contains a number of tyrosine residues and proline-rich sequences that can serve as docking sites for SH2 and SH3 containing proteins (see chapter 2). Other proteins that are able to function as docking/adapters in integrin-mediated signaling are p130Cas (see chapter 2), IRS-1 (Vuori and Ruoslahti 1994) and PI3-K (Chen and Guan 1994). Cytoskeletal protein themselves may play a role in assembling the transduction machinery. Paxillin and tensin, two proteins specifically recruited to focal adhesion sites, contain numerous domains mediating protein-protein interactions (see chapter 2). The role of cytoskeletal proteins in the signaling process is also indicated by the observation that cytochalasin D, an agent that causes depolimerization of filamentous actin and interferes with integrin signaling. This finding has been confirmed by several studies which showed that cytochalasin D treatment prevents p125Fak tyrosine phosphorylation (Lipfert et al 1992), MAPK activation (Chen et al 1994) and cell cycle progression (Bohmer et al 1996). Thus, polymerization of the actin cytoskeleton and assembly of the focal adhesion complexes in response to integrin-ligand interaction is required for correct triggering of the signaling response. A possible explanation for this effect is provided by the finding that assembly of the cytoskeleton causes inhibition of tyrosine phosphatases able to dephosphorylate p125Fak and paxillin (Defilippi et al 1995; Retta et al 1996).

The requirement for actin polymerization in integrin signaling also explains the necessity of cell spreading on the substratum for appropriate signaling. While p125Fak tyrosine phosphorylation can be triggered in absence of cell spreading (Defilippi et al 1994), more down stream events require cell spreading (Bohmer et al 1996).

The extracellular events necessary to induce integrin signaling involve receptor clustering in the plane of the plasma membrane as well as receptor occupancy by the ligand. Most of the intracellular responses can be triggered by antibody-induced integrin clustering at the cell surface, but some events require occupancy of the ligand binding site (Miyamoto et al 1995; Yamada and Miyamoto 1995). When integrin clustering is induced by antibodies that block ligand binding, actin, talin, vinculin, α-actinin, p125Fak and tensin cocluster with integrins. On the other hand, antibodies that do not block integrin-ligand binding induce coclustering of p125Fak and tensin, but not of other cytoskeletal proteins. Based on these and other data, a model has been proposed in which different steps in the signal transduction across the membrane can be postulated. While simple clustering of integrin in the plane of the plasma membrane can drive the organization of a transduction complex via tyrosine phosphorylation events, the organization of actin cytoskeleton requires interaction of the receptor with the matrix ligand (Yamada and Miyamoto 1995). The role of ligand binding in the integrin-actin cytoskeleton association can be explained by the release of the α subunit inhibition on the interaction of the β1 cytoplasmic domain with focal adhesion components (LaFlamme et al 1992; Briesewitz et al 1993; Ylanne et al 1993. See also above).

INTEGRIN-ASSOCIATED MEMBRANE PROTEINS ARE INVOLVED IN SIGNAL TRANSDUCTION

Recent data indicate that integrin associated proteins are important elements in signal transduction. These are membrane-proteins which are physically associated with integrins and modulate their activity. The first molecule of this class to be identified is a protein called IAP/CD47 (Integrin Associated Protein) (Brown et al 1990). IAP is a 50 kDa protein with multiple membrane-spanning domains (Lindberg et al 1993) that regulate both ligand binding and calcium signaling ability of αVβ3 (Brown et al 1990; Schwartz et al 1993). Several proteins characterized by four membrane-spanning segments, including CD9, CD63 and CD81, can also associate with integrins and modulate their signal transduction activity (Berditchevski et al 1995 and 1996; Shaw et al 1995a). Caveolin is an additional protein that can associate with integrins and couple them to other membrane receptors (Wei et al 1996) or to the adaptor protein Shc (Wary et al 1996). These proteins can be co-precipitated with integrin complexes and interact with the extracellular domains of α subunits (Berditchevski et al 1995; Wary et al 1996). Thus interaction of the integrin extracellular domain with these proteins in the plane of the plasma membrane can be of critical importance in determining their functional properties and signaling specificity.

REFERENCES

Altieri DC, Edgington TS (1988) A monoclonal antibody reacting with distinct adhesion molecules defines a transition in the functional state of the receptor CD11b/CD18 (Mac-1). J Immunol 141:2656-2660.

Altruda F, Cervella P, Tarone G, Botta C, Balzac F, Stefanuto G, Silengo L (1990) A human integrin β1 subunit with a unique cytoplasmic domain generated by alternative mRNA processing. Gene 95:261-216.

Arroyo AG, Garcia-Pardo A, Sanchez-Madrid F (1993) A high-affinity conformational state on VLA integrin heterodimers induced by an anti-β1 chain monoclonal antibody. J Biol Chem 268:9863-9868.

Balzac F, Belkin AM, Koteliansky VE, Balabanov YV, Altruda F, Silengo L, Tarone G (1993) Expression and functional analysis of a cytoplasmic domain variant of the β1 integrin subunit. J Cell Biol 121:171-178.

Balzac F, Retta SF, Albini A, Melchiorri A, Koteliansky VE, Geuna M, Silengo L, Tarone G (1994) Expression of β1B integrin isoform in CHO cells results in a dominant negative effect on cell adhesion and motility. J Cell Biol 127:557-565.

Baudoin C, van der Flier A, Borradori l, Sonnenberg A (1996) Genomic organization of the mouse β1 gene: conservation of the β1D, but not of the β1B and β1C integrin splice variants. Cell Adh Comm 4:1-11.

Belkin A, Pletjushkina O, Balzac F, Silengo L, Chi-Rosso G, Koteliansky V, Burridge K, Tarone G. Modulation of integrin function by alternative splicing: muscle β1D integrin reinforces the cytoskeletal-matrix link. Submitted.

Belkin AM, Zhidkova NI, Balzac F, Altruda F, Tomatis D, Maier A, Tarone G, Koteliansky VE, Burridge K (1996) B1D integrin displaces the β1A isoform in striated muscles: localization at junctional structures and signaling potential in non-muscle cells. J Cell Biol 132:211-226.

Bennet JS, Vilaire G (1979) Exposure of platelet fibrinogen receptors by ADP and epinephrine. J Clin Invest 64:1393-1401.

Berditchevski F, Bazzoni G, Hemler ME (1995) Specific association of CD63 with the VLA-3 and VLA-6 integrins. J Biol Chem 270:17784-17790.

Berditchevski F, Zutter MM, Hemler ME (1996) Characterization of novel complexes on the cell surface between integrins and proteins with 4 transmembrane domains (TM4 proteins). Mol Biol Cell 7:193-207.

Bohmer RM, Scharf E, Assoian RK (1996) Cytoskeletal integrity is required throughout the mitogen stimulation phase of the cell cycle and mediates the anchorage-dependent expression of cyclin D1. Mol Biol Cell 7:101-111.

Briesewitz R, Kern A, Marcantonio EE (1993) Ligand-dependent and -independent integrin focal contact localization: the role of the α chain cytoplasmic domain. Mol. Biol. Cell 4:593-604.

Brown E, Hooper L, Ho T, Gresham H (1990) Integrin-associated protein: a 50-kD plasma membrane antigen physically and functionally associated with integrins. J Cell Biol 111:2785-2794.

Burridge K, Fath K, Kelly T, Nuckolls G, Turner C (1988) Focal adhesions: transmembrane junctions between the extracellular matrix and the cytoskeleton. Annu Rev Cell Biol 4:487-525.

Calvete JJ, Henschen A, Gonzales-Rodriguez J (1991) Assignment of disulphide bonds in human platelet GPIIIa. A disulphide pattern for the β subunit of the integrin family. Biochem J 274:63-71.

Cattelino A, Longhi R, de Curtis I (1995) Differential distribution of two cytoplasmic variants of the α6β1 integrin laminin receptor in the ventral plasma membrane of embryonic fibroblasts. J Cell Sci 108:3067-3078.

Cepek KL, Shaw SK, Parker CM, Russell GJ, Morrow S, Rimm DL, Brenner MB (1994) Adhesin between epithelial cells and T lymphocytes mediated by E-cadherin and the αEβ7 integrin. Nature 372:190-193.

Charo IF, Nannizzi L, Phillips DR, Hsu MA, Scarborough RM (1991) Inhibition of fibrinogen binding to GP IIb-IIIa by a GP IIIa peptide. J Biol Chem 266:1415-1421.

Chen HC, Appeddu PA, Parsons JT, Hildebrand JD, Schaller MD Guan JL (1995) Interaction of focal adhesion kinase with cytoskeletal protein talin. J Biol Chem 270:16995-16999.

Chen HC, Guan JL (1994) Association of focal adhesion kinase with its potential substrate phosphatidylinositol 3-kinase. Proc Natl Acad Sci USA 91:10148-10152.

Chen Q, Kinch MS, Lin TS, Burridge K, Juliano RL (1994) Integrin-mediated cell adhesion activates mitogen-activated protein kinases. J Biol Chem 269:26602-26605.

Chen WT, Wang J, Hasegawa T, Yamada SS, Yamada KM (1986) Regulation of fibronectin receptor distribution by transformation, exogenous fibronectin, and synthetic peptides. J Cell Biol 103:1649-1661.

Chen YP, Djaffar I, Pidard D, Steiner B, Cieutat AM, Caen JP, Rosa JP (1992) Ser-752-Pro mutation in the cytoplasmic domain of integrin β3 subunit and defective activation of platelet integrin αIIb β3 (glycoprotein IIb-IIIa) in a variant of Glanzmann thrombasthenia. Proc Natl Acad Sci USA 89:10169-10173.

Clarke AS, Lotz MM, Mercurio AM (1994) A novel structural variant of the human β4 integrin cDNA. Cell Adhes Commun 2:1-6.

Collo G, Starr L, Quaranta V (1993) A new isoform of the laminin receptor integrin α7 β1 is developmentally regulated in skeletal muscle. J Biol Chem 268:19019-19024.

Conforti G, Zanetti A, Pasquali-Ronchetti I, Quaglino D Jr, Neyroz P, Dejana E (1990) Modulation of vitronectin receptor binding by membrane lipid composition. J Biol Chem 265:4011-4019.

Cooper HM, Tamura RN, Quaranta V (1991) The major laminin receptor of mouse embryonic stem cells is a novel isoform of the α6 β1 integrin. J Cell Biol 115:843-850.

Coppolino M, Leung-Hagesteijn C, Dedhar S, Wilkins J (1995) Inducible interaction of integrin α2 β1 with calreticulin. Dependence on the activation state of the integrin. J Biol Chem 270:23132-23138.

Crippes BA, Engleman VW, Settle SL, Delarco J, Ornberg RL, Helfrich MH, Horton MA, Nickols GA (1996) Antibody to beta3 integrin inhibits osteoclast-mediated bone resorption in the thyroparathyroidectomized rat. Endocrinology 137:918-924.

Crowe D, Chiu H, Fong S, Weissman, I (1994) Regulation of the avidity of integrin β7 by the β7 cytoplasmic domain. J Biol Chem 269:14111-14118.

D'Souza SE, Ginsberg MH, Burke TA, Plow EF (1990) The ligand binding site of the platelet integrin receptor GPIIb-IIIa is proximal to the second calcium binding domain of its α subunit. J Biol Chem 265:3440-3446.

Damsky CH, Knudsen KA, Bradley D, Buck CA, Horwitz AF (1985) Distribution of the cell substratum attachment (CSAT) antigen on myogenic and fibroblastic cells in culture. J Cell Biol 100:1528-1539.

Dana N, Fathallah DM, Arnaout MA (1991) Expression of a soluble and functional form of the human β2 integrin CD11b/CD18. Proc Natl Acad Sci USA 88:3106-3110.

Danilenko DM, Rossitto PV, van der Vieren, Trong HL, McDonough SP, Affolter VK, Moore PF (1995) A novel canine leukointegrin, adb2, is expressed by specific macrophage subpopulations in tissue and minor CD8+ lymphocyte subpopulation in peripheral blood. J Immunol 155:35-44.

Dedhar S (1989) Signal transduction via the β1 integrins is a required intermediate in interleukin-1β induction of alkaline phosphatase activity in human osteosarcoma cells. Exp Cell Res 183:207-214.

Defilippi P, Bozzo C, Volpe G, Romano G, Venturino M, Silengo M, Tarone G (1994) Integrin-mediated signal transduction in human endothelial cells, analysis of tyrosine phosphorylation events. Cell Adhes Commun 2:75-86.

Defilippi P, Retta SF, Olivo C, Palmieri M, Venturino M, Silengo L, Tarone G (1995) p125FAK tyrosine phosphorylation and focal adhesion assembly: studies with phosphotyrosine phosphatase inhibitors. Exp Cell Res 221:141-152.

Defilippi P, van Hinsbergh V, Bertolotto A, Rossino P, Silengo L, Tarone G (1991) Differential distribution and modulation of expression of α1/β1 integrin on human endothelial cells. J Cell Biol 114:855-863.

Delwel GO, Hogervorst F, Kuikman I, Paulsson M, Timpl R, Sonnenberg A (1993) Expression and function of the cytoplasmic variants of the integrin α6 subunit in transfected K562 cells. Activation-dependent adhesion and interaction with isoforms of laminin. J Biol Chem 268:25865-25875.

Delwel GO, Kuikman I, Sonnenberg A (1995) An alternatively spliced exon in the extracellular domain of the human α6 integrin subunit—functional analysis of the α6 integrin variants. Cell Adhes Commun 3:143-161.

Diamond MS, Alon R, Pakos CA, Quinn MT, Springer TA (1995) Heparin is an adhesive ligand for the leukocyte integrin Mac-1 (CD11b/CD18). J Cell Biol 130:1473-1482.

Diamond MS, Garcia-Aguilar J, Bickford JK, Corbi AL, Springer TA (1993) The I domain is a major recognition site on the leukocyte integrin Mac-1 (CD11b/CD18) for four distinct adhesion ligands. J Cell Biol 120:1031-1043.

Dike LE, Farmer SR (1988) Cell adhesion induces expression of growth-associated genes in suspension-arrested fibroblasts. Proc Natl Acad Sci USA 85:6792-6796.

Djaffar I, Chen YP, Creminon C, Maclouf J, Cieutat AM, Gayet O, Rosa JP (1994). A new alternative transcript encodes a 60 kDa truncated form of integrin β3. Biochem J 300:69-74.

Dowling J, Yu QC, Fuchs E (1996) Beta4 integrin is required for hemidesmosome formation, cell adhesion and cell survival. J Cell Biol 134:559-572.

Dustin ML, Springer TA (1989) T cell receptor crosslinking transiently stimulates adhesiveness through LFA-1. Nature 341:619-624.

Edgar D, Timpl R, Thoenen H (1984) The heparin-binding domain of laminin is responsible for its effects on neurite out-growth and neuronal survival. EMBO J 3:1463-1468.

Eierman DF, Johnson CE, Haskill JS (1989) Human monocyte inflammatory mediator gene expression is selectively regulated by adherence substrates. J Immunol 142:1970-1976.

El Ghmati SM, van Hoeyveld EM, van Strijp JAG, Ceuppens JL, Stevens EAM (1996) Identification of haptoglobin as an alternative ligand for CD11b/CD18. J Immunol 156:2542-2552.

Fässler R, Georges-Labouesse E, Hirsch E (1996) Genetic analysis of integrin function in mice. Curr Op Cell Biol 8:641-646.

Fässler R, Mayer M (1995) Consequences of lack of β1 integrin gene expression in mice. Genes Dev 9:1896-1908.

Faull RJ, Kovach NL, Harlan JM, Ginsberg MH (1993) Affinity modulation of integrin α5β1 regulation of the functional response by soluble fibronectin. J Cell Biol 121:155-162.

Ferrel, JE Martin GS (1989) Tyrosine-specific protein phosphorylation is regulated by glycoprotein IIb-IIIa in platelets. Proc Natl Acad Sci USA 86:2234-2238.

Folkman J, Moscona A (1978) Role of cell shape in growth control. Nature 273:345-349.

Fornaro M, Zheng DQ, Languino LR (1995) The novel structural motif Gln795-Gln802 in the integrin β1C cytoplasmic domain regulates cell proliferation. J Biol Chem 270:24666-9.

Frelinger AL, Du XP, Plow EF, Ginsberg MH (1991) Monoclonal antibodies to ligand-occupied conformers of integrin αIIb β3 (glycoprotein IIb-IIIa) alter receptor affinity, specificity, and function. J Biol Chem 266:17106-17111.

Gailit J, Ruoslahti E (1988) Regulation of the fibronectin receptor affinity by divalent cations. J Biol Chem 263:12927-12932.

Giancotti F, Tarone G, Damsky C, Knudsen K, Comoglio PM (1985) Cleavage of a 135K cell surface glycoprotein prevents cell adhesion to fibronectin-coated dishes. Exp Cell Res 156:182-190.

Giancotti FG, Comoglio PM, Tarone G (1986) A 135.000 molecular weight plasma membrane glycoprotein involved in fibronectin-mediated cell adhesion: immunofluorescence localization in normal and RSV-transformed fibroblasts. Exp Cell Res 163:47-62.

Ginsberg MH, Du X, Plow EF (1992) Inside-out integrin signalling. Curr Opin Cell Biol 4:766-771.

Guan JL, Trevithick JE, Hynes RO (1991) Fibronectin/integrin interaction induces tyrosine phosphorylation of a 120-kDa protein. Cell Regul 2:951-964.

Gulino D, Boudignon C, Zhang LY, Concord E, Rabiet MJ, Marguerie G (1992) Ca$^{(2+)}$-binding properties of the platelet glycoprotein IIb ligand-interacting domain. J Biol Chem 267:1001-1007.

Hannigan GE, Leung-Hagesteijn C, Fitz-Gibbon L, Coppolino MG, Radeva G, Filmus J, Bell JC, Dedhar S (1996) Regulation of cell adhesion and anchorage-dependent growth by a new β1-integrin-linked protein kinase. Nature 379:91-96.

Harlan JM (1993) Leukocyte adhesion deficiency syndrome: insights into the molecular basis of leukocyte emigration. Clin. Immunol. Immunopathol. 67:S16-S24.

Hemler ME (1990) VLA proteins in the integrin family: structures, functions and their role on leukocytes. Annu Rev Immunol 8:365-400.

Hibbs ML, Xu H, Stacker SA, Springer TA (1991) Regulation of adhesion of ICAM-1 by the cytoplasmic domain of LFA-1 integrin β subunit. Science 251:1611-1613.

Hogervorst F, Admiraal LG, Niessen C, Kuikman I, Janssen H, Daams H, Sonnenberg A (1993a) Biochemical characterization and tissue distribution of the A and B variants of the integrin α6 subunit. J Cell Biol 121:179-191.

Hogervorst F, Kuikman I, Noteboom E, Sonnenberg A (1993b) The role of phosphorylation in activation of the α6Aβ1 laminin receptor. J Biol Chem 268:18427-30.

Hogervorst F, Kuikman I, van Kessel AG, Sonnenberg A (1991) Molecular cloning of the human α6 integrin subunit. Alternative splicing of α6 mRNA and chromosomal localization of the α6 and β4 genes. Eur J Biochem 199:425-433.

Hogervorst F, Kuikman I, von dem Borne AE, Sonnenberg A (1990) Cloning and sequence analysis of beta-4 cDNA: an integrin subunit that contains a unique 118 kd cytoplasmic domain. EMBO J 9:765-770.

Horwitz A, Duggan K, Buck C, Beckerle MC, Burridge K (1986) Interaction of plasma membrane fibronectin receptor with talin: a transmembrane linkage. Nature 320:531-533.

Hughes PE, Diaz-Gonzalez F, Leong L, Wu C, McDonald JA, Shattil SJ, Ginsberg MH (1996) Breaking the integrin hinge. J Biol Chem 271:6571-6574.

Hughes PE, O'Toole TE, Ylanne J, Shattil SJ, Ginsberg MH (1995) The conserved membrane-proximal region of an integrin cytoplasmic domain specifies ligand binding affinity. J Biol Chem 270:12411-12417.

Huhtala P. Humphries MJ. McCarthy JB. Tremble PM. Werb Z. Damsky CH (1995) Cooperative signaling by α5 β1 and α4 β1 integrins regulates metalloproteinase gene expression in fibroblasts adhering to fibronectin. J Cell Biol 129:867-879.

Hynes RO (1987) Integrins: a family of cell surface receptors. Cell 48:549-554.

Ignotz RA, Massague J (1987) Cell adhesion protein receptors as targets for transforming growth factor-β action. Cell 51:189-197.

Ingalls RR, Golenbock DT, (1995) CD11c/CD18, a transmembrane signalling receptor for lipopolysaccharide. J Exp Med 181:1473-1479.

Inghirami G, Grignani F, Sternas L, Lombardi L, Knowles DM, Dalla-Favera R (1990) Down-regulation of LFA-1 adhesion receptors by c-myc oncogene in human B lymphoblastoid cells. Science 250:682-686.

Isberg RR, Leong JM (1990) Multiple β1 chain integrins are receptors for invasin, a protein that promotes bacterial penetration into mammalian cells. Cell 60:861-71.

Jiang Y, Beller DI, Frendl G, Graves DT (1992) Monocyte chemoattractant protein-1 regulates adhesion molecule expression and cytokine production in human monocytes. J Immunol 148:2423-2428.

Jockusch BM, Bubeck P, Giehl K, Kroemker M, Moschner J, Rothkegel M, Rüdiger M, Schlüter K, Stanke G, Winkler J (1995) The molecular architecture of focal adhesions. Annu Rev Cell Dev Biol 11:379-416.

Kamata T, Takada Y (1994) Direct binding of collagen to the I domain of integrin α2 β1 (VLA-2, CD49b/CD29) in a divalent cation-independent manner. J Biol Chem 269:26006-26010.

Kawasaki H, Kretsinger RH (1995) Calcium-binding proteins 1: EF hands. Protein Profile 2:297-490.

Keizer GD, Visser W, Vliem M, Figdor CG (1988) A monoclonal antibody (NKI-L16) directed against a unique epitope on the alpha-chain of human leukocyte function-asociated antigen 1 induces homotypic cell-cell interaction. J Immunol 140:1393-1400.

Kirchhofer D, Gailit J, Ruoslahti E, Grzesiak J, Pierschbacher MD (1990) Cation-dependent changes in the binding specificity of the platelet receptor GPIIb/IIIa. J Biol Chem 265:18525-18530.

Kitazawa S, Ross FP, McHugh K, Teitelbaum SL (1995) Interleukin-4 induces expression of the integrin αv β3 via transactivation of the β3 gene. J Biol Chem 270:4115-4120.

Klemke RL, Yebra M, Bayna EM, Cheresh DA (1994) Receptor tyrosine kinase signaling required for integrin αv β5-directed cell motility but not adhesion on vitronectin. J Cell Biol. 127:859-866.

Kolanus W, Nagel W, Schiller B, Zeitlmann L, Godar S, Stockinger H, Seed B (1996) αLβ2 integrin/LFA-1 binding to ICAM-1 induced by cytoadhesin-1, a cytoplasmic regulatory molecule. Cell 86:233-242.

Kovach NL, Carlos TM, Yee E, Harlan JM (1992) A monoclonal antibody to β1 integrin (CD29) stimulates VLA-dependent adherence of leukocytes to human umbilical vein endothelial cells and matrix components. J Cell Biol 116:499-509.

LaFlamme SE, Akiyama SK, Yamada KM (1992) Regulation of fibronectin receptor distribution. J Cell Biol 117:437-442.

Languino LR, Ruoslahti E (1992) An alternative form of the integrin β1 subunit with a variant cytoplasmic domain. J Biol Chem 267:7116-7120.

Laudanna C. Campbell JJ. Butcher EC (1996) Role of Rho in chemoattractant-activated leukocyte adhesion through integrins. Science 271:981-983.

Lee JO, Rieu P, Arnaout MA, Liddington R (1995) Crystal structure of the A domain from the α subunit of integrin CR3 (CD11b/CD18). Cell 80:631-638.

Lenter M, Uhlig H, Hamann A, Jeno P, Imhof B, Vestweber D (1993) A monoclonal antibody against an activation epitope on mouse integrin chain β1 blocks adhesion of lymphocytes to the endothelial integrin α6 β1. Proc Natl Acad Sci USA 90:9051-9055.

Leptin M (1986) The fibronectin receptor family. Nature 321:728.

Leung-Hagesteijn CY, Milankov K, Michalak M, Wilkins J, Dedhar S (1994) Cell attachment to extracellular matrix substrates is inhibited upon downregulation of expression of calreticulin, an intracellular integrin alpha-subunit-binding protein. J Cell Sci 107:589-600.

Liaw L, Skinner MP, Raines EW, Ross R, Cheresh DA, Schwartz SM, Giachelli CM (1995) The adhesive and migratory effects of osteopontin are mediated via distinct cell surface integrins. Role of αv β3 in smooth muscle cell migration to osteopontin in vitro. J Clin Invest 95:713-724.

Lindberg FP, Gresham HD, Schwarz E, Brown EJ (1993) Molecular cloning of integrin-associated protein: an immunoglobulin family member with multiple membrane-spanning domains implicated in αv β3-dependent ligand binding. J Cell Biol 123:485-496.

Lipfert L, Haimovich B, Schaller MD, Cobb BS, Parsons JT, Brugge JS (1992) Integrin-dependent phosporylation and activation of the protein tyrosine kinase pp125FAK in platelets. J Cell Biol 119:905-912.

Loftus JC, O'Toole TE, Plow EF, Glass A, Frelinger AL, Ginsberg MH (1990) A β3 integrin mutation abolishes ligand binding and alters divalent cation-dependent conformation. Science 249:915-918.

MacKrell A, Blumberg B, Haynes SR, Fessler JH (1988) The lethal myospheroid gene of drosophila encodes a membrane protein homologous to vertebrate integrin β subunits. Proc Natl Acad Sci USA 85:2633-2637.

Mainiero F, Pepe A, Wary KK, Spinardi L, Mohammadi M, Schlessinger J, Giancotti FG (1995) Signal transduction by the α6 β4 integrin: distinct β4 subunit sites mediate recruitment of Shc/Grb2 and association with the cytoskeleton of hemidesmosomes EMBO J 14:4470-4481.

Marcantonio EE, Guan JL, Trevithick JE, Hynes RO (1990) Mapping of the functional determinants of the integrin β1 cytoplasmic domain by site-directed mutagenesis. Cell Regul 1:597-604.

Martin PT, Kaufman SJ, Kramer RH, Sanes JR (1996) Synaptic integrins in developing, adult, and mutant muscle: selective association of alpha1, alpha7A, and alpha7B integrins with the neuromuscular junction. Dev. Biol. 174:125-139.

Menko AS, Boettiger D (1987) Occupation of the extracellular matrix receptor, integrin, is a central point for myogenic differentiation. Cell 51:51-57.

Meredith J Jr, Takada Y, Fornaro M, Languino LR, Schwartz MA (1995) Inhibition of cell cycle progression by the alternatively spliced integrin β1C. Science 269:1570-1572.

Michishita M, Videm V, Arnaout MA (1993) A novel divalent cation-binding site in the A domain of the β2 integrin CR3 (CD11b/CD18) is essential for ligand binding. Cell 72:857-867.

Miyamoto S, Akiyama SK, Yamada KM (1995) Synergistic roles for receptor occupancy and aggregation in integrin transmembrane function. Science 267:883-885.

Mould AP, Garrat AN, Askari JA, Akiyama SK, Humphries MJ (1995) Identification of a novel anti-integrin monoclonal antibody that recognizes a ligand-induced binding site epitope on beta1 subunit. FEBS Lett 363:118-122.

Nermut MV, Green NM, Eason P, Yamada SS, Yamada KM (1988) Electron microscopy and structural model of human fibronectin receptor. EMBO J 7:4093-4099.

Neugebauer KM, Reichardt LF (1991) Cell-surface regulation of β1-integrin activity on developing retinal neurons. Nature 350:68-71.

O'Toole TE, Katagiri Y, Faull RJ, Peter K, Tamura R, Quaranta V, Loftus JC, Shattil SJ, Ginsberg MH (1994) Integrin cytoplasmic domains mediate inside-out signal transduction. J Cell Biol 124:1047-1059.

O'Toole TE, Mandelman D, Forsyth J, Shattil SJ, Plow EF, Ginsberg MH (1991)

Modulation of the affinity of integrin αIIb β3 (GPIIb-IIIa) by the cytoplasmic domain of αIIb. Science 254:845-847.

O'Toole TE, Ylanne J, Culley BM (1995) Regulation of integrin affinity states through an NPXY motif in the β subunit cytoplasmic domain. J Biol Chem 270:8553-8558.

Olchowy TW. Bochsler PN. Welborn MG (1994) Clinicopathological findings in a Holstein calf with peripheral leukocytosis and leukocyte adhesion deficiency. Canad Vet J 35:242-243.

Otey CA, Pavalko FM, Burridge K (1990) An interaction between alpha-actinin and the β1 integrin subunit in vitro. J Cell Biol 111:721-729.

Otey CA, Vasquez GB, Burridge K, Erickson BW (1993) Mapping of the alpha-actinin binding site within the beta1 integrin cytoplasmic domain. J Biol Chem 268:21193-21197.

Pardi R, Bender JR, Dettori C, Giannazza E, Engelman EG (1989) Heterogeneous distribution and transmembrane signaling properties of lymphocyte function-associated antigen (LFA-1) in human lymphocyte subsets. J Immunol 143:3157-3166.

Pardi R, Inverardi L, Rugarli C, Bender J. (1992) Antigen receptor complex stimulation triggers protein kinase C dependent CD11a/CD18-cytoskeleton association in T lymphocytes. J Cell Biol 116:1211-1220.

Pavalko FM, LaRoche SM (1993) Activation of human neutrophils induces an interaction between the integrin β2-subunit (CD18) and the actin binding protein alpha-actinin. J Immunol 151:3795-3807.

Perutelli P, Mori PG (1992) Biochemical and molecular basis of Glanzmann's thrombasthenia. Haematologica 77:421-426.

Piali L, Hammel P, Uherek C, Bachmann F, Gisler RH, Dunon D, Imhof BA (1995) CD31/PECAM-1 is a ligand for αv β3 integrin involved in adhesion of leukocytes to endothelium. J Cell Biol 130:451-460.

Pierschbacher MD, Ruoslahti E (1984) Cell attachment activity of fibronectin can be duplicated by small synthetic fragments of the molecule. Nature 309:30-33.

Plantefaber LC, Hynes RO (1989) Changes in integrin receptors on oncogenically transformed cells. Cell 56:281-290.

Pytela R, Pierschbacher MD, Argraves S, Suzuki S, Ruoslahti E (1987) Arg-Gly-Asp adhesion receptors. Meth Enzymol 144:475- 489.

Pytela R, Pierschbacher MD, Ruoslahti E (1985a) Identification and isolation of a 140 kd cell surface glycoprotein with properties expected of a fibronectin receptor. Cell 40:191-198.

Pytela R, Pierschbacher MD, Ruoslahti E (1985b) A 125/115-kDa cell surface receptor specific for vitronectin interacts with the arginine-glycine-aspartic acid adhesion sequence derived from fibronectin. Proc Natl Acad Sci USA 82:5766-5770.

Randi AM, Hogg N (1994) I domain of β2 integrin lymphocyte function-associated antigen-1 contains a binding site for ligand intercellular adhesion molecule-1. J Biol Chem 269:12395-12398.

Relman D, Tuomanen E, Falkow S, Golenbock DT, Saukkonen K, Wright SD (1990) Recognition of a bacterial adhesion by an integrin: macrophage CR3 (αM β2, CD11b/CD18) binds filamentous hemagglutinin of Bordetella pertussis. Cell 61:1375-82.

Reszka AA, Hayashi RA, Horwitz AF (1992). Identification of amino acid sequences in the integrin β1 cytoplasmic domain implicated in cytoskeletal association. J Cell Biol 117:1321-1330.

Retta SF, Balzac F, Fässler R, Humphries MJ, Silengo L, Tarone G. The cytoplasmic domain of the β1B integrin isoform negatively regulates ligand binding and fibronectin matrix assembly. Submitted.

Retta SF, Barry ST, Critchley DR, Defilippi P, Silengo L, Tarone G (1996) Focal adhesion and stress fibre formation is regulated by tyrosine phosphatase activity. Exp Cell Res 229:307-317.

Rojiani MV, Finlay BB, Gray V, Dedhar S (1991) In vitro interaction of a polypeptide homologous to human Ro/SS-A antigen (calreticulin) with a highly conserved amino acid sequence in the cytoplasmic domain of integrin α subunits. Biochemistry 30:9859-9866.

Rossino P, Gavazzi I, Timpl R, Aumailley M, Abbadini M, Giancotti F, Silengo L, Marchisio PC, Tarone G (1990) Nerve growth factor induces increased expression of a laminin-binding integrin in rat pheochromocytoma PC12 cells. Exp Cell Res 189:100-108.

Ruoslahti E, Pierschbacher MD (1987) New perspectives in cell adhesion: RGD and integrins. Science 238:491-497.

Ruppert M, Aigner S, Hubbe M, Yagita H, Altevogt P (1995) The L1 adhesion molecule is a cellular ligand for VLA-5. J Cell Biol 131:1881-1891.

Santala P, Heino J (1991) Regulation of integrin-type cell adhesion receptors by cytokines. J Biol Chem 266:23505-23509.

Sastry SK, Horwitz AF (1993) Integrin cytoplasmic domains: mediators of cytoskeletal linkages and extra- and intracellular initiated transmembrane signaling. Curr Opin Cell Biol 5:819-831.

Schaller MD, Otey CA, Hildebrand JD, Parsons JT (1995) Focal adhesion kinase and paxillin bind to peptides mimicking β integrin cytoplasmic domains. J Cell Biol 130:1181-1187.

Schwartz MA, Both G, Lechene C (1989) Effect of cell spreading on cytoplasmic pH in normal and transformed fibroblasts. Proc Natl Acad Sci USA 86:4525-4529.

Schwartz MA, Brown EJ, Fazeli B (1993) A 50 kDa integrin associated protein is required for integrin-regulated calcium entry in endothelial cells. J Biol Chem 268:19931-19934.

Sharma CP, Ezzell RM, Arnaout MA (1995) Direct interaction of filamin (ABP-280) with the β2 integrin subunit CD18. J Immunol 154:3641-3470.

Shattil SJ, Cunningham M, Wiedmer T, Zhao J, Sims PJ, Brass LF (1992) Regulation of glycoprotein IIb-IIIa receptor function studied with platelets permeabilized by the pore-forming complement proteins C5b-9. J Biol Chem 267:18424-18431.

Shattil SJ, O'Toole T, Eigenthaler M, Thon V, Williams M, Babior BM, Ginsberg MH (1995) B3-endonexin, a novel polypeptide that interacts specifically with the cytoplasmic tail of the integrin β3 subunit. J Cell Biol 131:807-816.

Shaw AR, Domanska A, Mak A, Gilchrist A, Dobler K, Visser L, Poppema S, Fliegel L, Letarte M, Willett BJ (1995a) Ectopic expression of human and feline CD9 in a human B cell line confers β1 integrin-dependent motility on fibronectin and laminin substrates and enhanced tyrosine phosphorylation. J Biol Chem 270:24092-24099.

Shaw LM, Meisser JM Mercurio AM (1990) The activation dependent adhesion of macrophages to laminin involves cytoskeletal anchoring and phosphorylation of the a6b1 integrin. J Cell Biol 110:2167-2174.

Shaw LM, Mercurio AM (1993) Regulation of α6 β1 integrin laminin receptor function by the cytoplasmic domain of the α6 subunit. J Cell Biol 123:1017-1025.

Shaw LM, Mercurio AM (1994) Regulation of cellular interactions with laminin by integrin cytoplasmic domains: the A and B structural variants of the α6 β1 integrin differentially modulate the adhesive strength, morphology, and migration of macrophages. Mol Biol Cell 5:679-690.

Shaw LM, Turner CE, Mercurio AM (1995b) The α6Aβ1 and α6Bβ1 integrin variants signal differences in the tyrosine phosphorylation of paxillin and other proteins. J Biol Chem 270:23648-23652.

Smith JW, Cheresh DA (1990) Integrin (αv β3)-ligand interaction. Identification of a heterodimeric RGD binding site on the vitronectin receptor. J Biol Chem 265:2168-2172.

Socinski MA, Cannistra SA, Sullivan R, Elias A, Antman K, Schnipper L, Griffin JD (1988) Granulocyte-macrophage colony-stimulating factor induces the expression of the CD11b surface adhesion molecule on human granulocytes in vivo. Blood 72:691-697.

Solowska J, Guan JL, Marcantonio EE, Trevithick JE, Buck CA, Hynes RO (1989) Expression of normal and mutant avian integrin subunits in rodent cells. J Cell Biol 109:853-861.

Solursh M, Jensen KL, Zanetti NC, Linsenmayer TF, Reiter RS (1984) Extracellular matrix mediates epithelial effects on chondrogenesis in vitro. Dev Biol 105:451-457.

Stephens LE, Sutherland AE, Kimanskaya IV, Andrieux A, Meneses J, Pedersen RA, Damsky CH (1995) Deletion of β1 integrin in mice results in inner cell mass failure and peri-implantation lethality. Genes Dev 9:1883-1895.

Suzuki S, Naitoh Y (1990) Amino acid sequence of a novel integrin β4 subunit and primary expression of the mRNA in epithelial cells. EMBO J 9:757-763.

Tamura R, Rozzo C, Starr L, Chambers J, Reichard L, Quaranta V (1990) Epithelial integrin α6β4 primary structure of α6 and variant forms of β4. J Cell Biol 111:1593-1604.

Tamura RN, Cooper HM, Collo G, Quaranta V (1991) Cell type-specific integrin variants with alternative αchain cytoplasmic domains. Proc Natl Acad Sci USA 88:10183-10187.

Tapley P, Horwitz A, Buck C, Duggan K, Rohrschneider L (1989) Integrins isolated from Rous sarcoma virus-transformed chicken embryo fibroblasts. Oncogene 4:325-33.

Tarone G, Galetto G, Prat M, Comoglio PM (1982) Cell surface molecules and fibronectin mediated cell adhesion: effect of proteolytic digestion of membrane proteins. J Cell Biol 39:179-185.

Thorsteinsdottir S, Roelen BA, Freund E, Gaspar AC, Sonnenberg A, Mummery CL (1995) Expression patterns of laminin receptor splice variants α6A β1 and α6B β1 suggest different roles in mouse development. Dev Dynam 204:240-258.

Tiisala S, Paavonen T, Renkonen R (1995) αEβ7 and α4β7 integrins associated with intraepithelial and mucosal homing, are expressed on macrophages. Eur J Immunol 25:411-417.

Timpl R, Brown JC (1994) The laminins. Matrix Biol 14:275-281.

van der Flier A, Kuikman I, Baudoin C, van der Neut R, Sonnenberg A (1995) A novel β1 integrin isoform produced by alternative splicing: unique expression in cardiac and skeletal muscle. FEBS Lett 369:340-344.

van der Neut R, Krimpenfort P, Calafat J, Niessen CM, Sonnenberg A (1996) Epithelial detachment due to absence of hemidesmosomes in integrin β4 null mice. Nature Genet 13:366-369.

van der Vieren M, Trong HL, Wood CL, Moore PF, St John T, Staunton DE, Gallatin WM (1995) A novel leukointegrin, adb2, binds preferentially to ICAM-3. Immunity 3:638-690.

van Kooyk Y, Weder P, Heije K, de Waal Malefijt R, Figdor CG (1993) Role of intracellular Ca2+ levels in the regulation of CD11a/CD18 mediated cell adhesion. Cell Adhes Commun 1:21-32.

van Kooyk Y, Weder P, Heije K, Figdor CG (1994) Extracellular Ca^{2+} modulates leukocyte function-associated antigen-1 cell surface distribution on T lymphocytes and consequently affects cell adhesion. J Cell Biol 124:1061-1070.

van Kooyk Y, Weder P, Hogervorst F, Verhoeven AJ, van Seventer G, te Velde AA, Borst J, Keizer GD, Figdor CG (1991) Activation of LFA-1 through a Ca^{2+}-dependent epitope stimulates lymphocyte adhesion. J Cell Biol 112:345-354.

van Kuppevelt TH, Languino LR, Gailit JO, Suzuki S, Ruoslahti E (1989) An alternative cytoplasmic domain of the integrin β3 subunit. Proc Natl Acad Sci USA 86:5415-5418.

Vellig T, Collo G, Sorokin L, Durbeej M, Zhang H, Gullberg D (1996) Distinct α7Aβ1 and α7Bβ1 expression patterns during mouse development: α7A is restricted to skeletal muscle but α7B is expressed in striated muscle, vasculature and nervous system. Dev Dynam, in press.

Vidal F, Aberdam D, Miquel C, Christiano AM, Pulkkinen L, Uitto J, Ortonne JP, Meneguzzi G (1995) Integrin β4 mutations associated with junctional epidermolysis bullosa with pyloric atresia. Nature Genet 10:229-234.

Vuori K, Ruoslahti E (1994) Association of insulin receptor substrate-1 with integrins. Science 266:1576-1578.

Wardlaw AJ, Hibbs ML, Stacker SA, Springer TA (1990) Distinct mutations in two patients with leukocyte adhesion deficiency and their functional correlates. J Exp Med 172:335-345.

Wary KK, Mainiero F, Isakoff SJ, Marcantonio E, Giancotti FG (1996) The adaptor protein Shc couples a class of integrins to the control of cell cycle progression. Cell 87:733-744.

Wei Y, Lukashev M, Simon SI, Bodary SC, Rosenberg S, Doyle MV, Chapman HA (1996) Regulation of integrin function by the urokinase receptor. Science 273:1551-1554.

Wennenberg K, Lohikangas L, Gullberg D, Plaff M, Johansson S, Fässler R (1996) β1 integrin-dependent and independent polymerization of fibronectin. J Cell Biol 132:227-238.

Xu J, Clark RAF (1996) Extracellular matrix alters PDGF regulation of fibroblast integrins. J Cell Biol 132:239-249.

Yamada KM, Miyamoto S (1995) Integrin transmembrane signaling and cytoskeletal control. Curr Opin Cell Biol 7:681-689

Yang JT, Raybum H, Hynes RO (1993) Embryonic mesodermal defects in α5 integrin deficient mice. Development 119:1093-1105.

Ylanne J, ChenY, O' Toole TE, LOftus JC, Takada Y, Ginsberg MH (1993) Distinct functions of integrin and subunit cytoplasmic domains in cell spreading and formation of focal adhesions. J Cell Biol 122:223-233.

Ylanne J, Huuskonen J, O'Toole TE, Ginsberg MH, Virtanen I, Gahmberg CG (1995) Mutation of the cytoplasmic domain of the integrin β3 subunit. Differential effects on cell spreading, recruitment to adhesion plaques, endocytosis, and phagocytosis. J Biol Chem 270:9550-9557.

Zhang Z, Morla AO, Vuori K, Bauer JS. Juliano RL, Ruoslahti E (1993) The αv β1 integrin functions as a fibronectin receptor but does not support fibronectin matrix assembly and cell migration on fibronectin. J Cell Biol 122:235-242.

Zhang Z, Vuori K, Wang H, Reed JC, Ruoslahti E (1996) Integrin activation by R-ras. Cell 85:61-69.

Zhidkova NI, Belkin AM, Mayne R (1995) Novel isoform of β1 integrin expressed in skeletal and cardiac muscle. Biochem Biophys Res Commun 214:279-85.

Zhou L, Lee DH, Plescia J, Lau CY, Altieri DC (1994) Differential ligand binding specificities of recombinant CD11b/CD18 integrin I-domain. J Biol Chem 269:17075-17079.

Ziober BL, Vu MP, Waleh N, Crawford J, Lin CS, Kramer RH (1993) Alternative extracellular and cytoplasmic domains of the integrin α7 subunit are differentially expressed during development. J Biol Chem 268:26773-26783.

CHAPTER 2

Activation of Tyrosine Kinases

PROTEIN TYROSINE KINASE FAMILIES

Tyrosine phosphorylation of cellular proteins was originally identified as a signaling response triggered by growth factors and oncogenes. This event accounts for less than 0.1% of the total intracellular phosphorylation and is the result of the balance of protein tyrosine kinases (PTKs) and protein tyrosine phosphatases (PTPases).

PTKs can be divided into receptor tyrosine kinases (RTKs), which normally act as receptors for growth factors, and into soluble cytosolic tyrosine kinases (Van der Geer et al 1994). Cytosolic PTKs usually are functionally coupled to receptors devoid of PTK activity, such as receptors for cytokines, antigens and many hormones (Ziemiecki et al 1994). These receptors compensate for their lack of an intrinsic PTK activity by recruiting and/or activating cytosolic PTKs. The same is true for the extracellular matrix receptors, which have no enzymatic activity and are coupled with cytosolic PTKs to transduce signals.

The cytosolic PTKs can be divided into four major families: Fak, Src, Zap70/Syk and Jak. Each family is composed by distinct components sharing a high degree of structural similarity, with common domain architecture and regulatory mechanisms (Ziemiecki et al 1994). The cytosolic PTKs implicated in integrin-mediated adhesion include members of Fak, Src and Zap70/Syk families. The Jak family is involved in cytokine receptors signaling (Ihle and Kerr, 1995) and at present there are no indications of their involvement in integrin signal transduction. In the following paragraphs, we will discuss separately the Fak, Src and Zap70/Syk kinases.

ADHESION-INDUCED TYROSINE PHOSPHORYLATION

Immunofluorescence experiments with anti-phosphotyrosine antibodies showed that most of the tyrosine phosphorylated proteins are localized at cell-substratum and cell-cell contacts in both normal and transformed cells (Maher, 1985), suggesting that the membrane-cytoskeleton interaction is crucially regulated by these events. Tyrosine phosphorylation of cellular proteins is a primary response to integrin stimulation. The first evidence that integrin occupancy by ligands induces tyrosine phosphorylation came from studies of the fibrinogen receptor αIIbβ3 in platelets (Ferrell and Martin, 1989; Nakamura and Yakamura, 1989; Golden et al 1990). Further work showed that interaction of normal or transformed cells with several matrix proteins or integrin receptor clustering by specific antibodies triggers tyrosine phosphorylation of intracellular proteins (Fig. 2.1).

Signal Transduction by Integrins, by Paola Defilippi, Angela Gismondi, Angela Santoni and Guido Tarone. © 1997 Landes Bioscience.

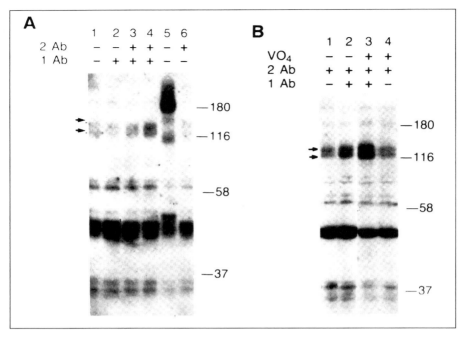

Fig. 2.1. Effect of integrin clustering on tyrosine phosphorylation in a human carcinoma cell line. (A) KB cells were incubated on ice for 30 min in presence of a 1:20 dilution of anti-integrin β1 subunit antibody (1 Ab). The cells were washed and then incubated in the presence of a 1:100 dilution of goat anti-rat IgG (2 Ab) at 37°C for 2 min (lane 3) or 10 min (lane 4). The cells in lane 5 were incubated for 2 min at 37°C in the presence of 300 ng of epidermal growth factor per ml. (B) KB cells were incubated with or without anti-integrin β1 antibody, as in A, and were then incubated at 37°C for 10 min with a 1:100 dilution of goat anti-rat IgG. Some dishes were incubated in the presence of 100 μM sodium vanadate. Following the incubations, the cells were analyzed for phosphotyrosine-containing proteins by western blot. The presence (+) or absence (-) of anti-integrin β1 antibody (1 Ab), goat anti-rat IgG (2 Ab) and sodium vanadate (VO₄) is as indicated. Positions of molecular mass markers (in kDa) and the p130 complex (double arrows) are indicated. With permission from Kornberg L et al Proc Natl Acad Sci USA 1992; 88:8392-8396.

Several components that are tyrosine phosphorylated in response to integrin stimulation have been identified. These include Fak and Src family members, p72Syk, p130Cas, paxillin, tensin, cortactin, PI-3 kinase, Shc and MAPK (see below and chapter 5). In addition a number of unidentified proteins have been described which are tyrosine phosphorylated in response to integrin-mediated adhesion in leukocytes (Nojima et al 1992; Freedman et al 1993; Sanchez-Mateos et al 1993; Brando and Shevach 1995; Gismondi et al 1995; Manie et al 1996) and in platelets (Huang et al 1993).

Fak FAMILY KINASES: STRUCTURE, EXPRESSION AND DISTRIBUTION

This family consists of cystolic PTKs localized in focal adhesions and includes a p125Fak of an apparent molecular weight of 116 kDa (Schaller et al 1992; Hanks et al 1992); and Pyk2 of 112 kDa (Lev et al 1995). Cakβ and Raftk kinases, originally described as unique molecules belonging to this family (Sasaki et al 1995; Avraham et al 1995), are indeed identical to Pyk2 (Hiregowdara et al 1997).

The cDNAs encoding p125Fak have been isolated from chicken, mouse, human and *Xenopus* (Schaller et al 1992; Hanks et al 1992; Andrè and Becker-Andrè, 1993; Whitney et al 1993; Zhang et al 1995a) and the analysis of the sequence shows a high level of homology between the different species, reaching 95% identity at the amino acid level.

The molecule is composed by the central kinase catalytic domain flanked by large aminoterminal and carboxyterminal sequences. The aminoterminal and the carboxyterminal domains have no homologies with other identified proteins and do not contain SH2 or SH3 domains. The aminoterminal and carboxyterminal domains contain several tyrosine residues and a number of structural features that are likely to be important in mediating physical association with signaling components and cytoskeletal proteins.

The aminoterminal domain of p125Fak contains the tyrosine residue 397, which has been identified as the primary site of tyrosine phosphorylation (Schaller et al 1994; Eide et al 1995). The region including amino acid residues 31-376 interacts in vitro with peptides reproducing the cytoplasmic domain of β1, β2 and β3 integrins (Schaller et al 1995).

The catalytic domain is only 31-41% identical to the catalytic domains of other PTKs; it contains the ATP-binding site and three residues (Lys-429, Glu-446 and Asp-539) known to interact with the γ-phosphate group of bound ATP. In addition this domain contains two tyrosine residues, Tyr-576 and Tyr-577, whose phosphorylation regulates the kinase activity of the molecule (Calalb et al 1995).

The carboxyterminal domain is characterized by two proline-rich domains (amino acid residues 712-721 and 875-884) that can be responsible for binding of SH3 domain-containing proteins (Guinebault et al 1995; Polte and Hanks, 1995; Hildebrand et al 1996). The Tyr-925 when phosphorylated can bind SH2 domain of Grb-2 (Schlaepfer et al 1994). In addition, the carboxyterminal domain contains the *focal adhesion targeting* (FAT) sequence, responsible for the lo-

calization of p125Fak to the focal adhesions (Hildebrand et al 1993), as well as binding sites for talin, paxillin, p130Cas and Graf-2 (Schaller and Parson, 1995; Hildebrand et al 1995; Chen et al 1995; Polte and Hanks, 1995; Hildebrand et al 1996). Fig. 2.2 presents a diagram illustrating the major structural features and binding sites in the p125Fak molecule.

p125Fak is widely expressed in the majority of established cell lines (Kornberg et al 1992; Burridge et al 1992; Hanks et al 1992), in platelets (Lipfert et al 1992), embryonic stem cells, lymphocytes and erythroid cells (Choi et al 1993; Nojima et al 1995a), endothelial, neuronal and vascular smooth muscle cells (Defilippi et al 1994; Romer et al 1994; Bozzo et al 1994; Polte et al 1994), keratinocytes (Gates et al 1994) and melanocytes (Scott and Liang, 1995). In contrast, macrophages, mast cells and monocytes express only trace levels of p125Fak mRNA (Choi et al 1993). In vivo p125Fak is highly expressed at the earlier stages of development (Turner et al 1993; Polte et al 1994). Specific localization in the walls of new blood vessels has been reported in mouse embryos (Polte et al 1994), while in adult mice the expression is less restricted (Hanks et al 1992; Polte et al 1994). In Xenopus, p125Fak is expressed in embryos during tailbud and tadpole stages (Zhang et al 1995a).

A truncated form of p125Fak, p41/43FRNK (Fak-related nonkinase) has been described consisting in the carboxyterminal noncatalytic domain of the p125Fak. This molecule, coded for by a specific transcript, is expressed in chicken embryo fibroblasts, where it localizes in focal adhesions (Schaller et al 1993) and behaves as a dominant negative regulator of p125Fak (see below) (Richardson and Parsons, 1996).

In human, Fak variants have been reported containing an insertion of 229 amino acids in the kinase domain and a brain-specific form lacking the aminoterminal 57 amino acids (Andrè and Becker-Andrè, 1993).

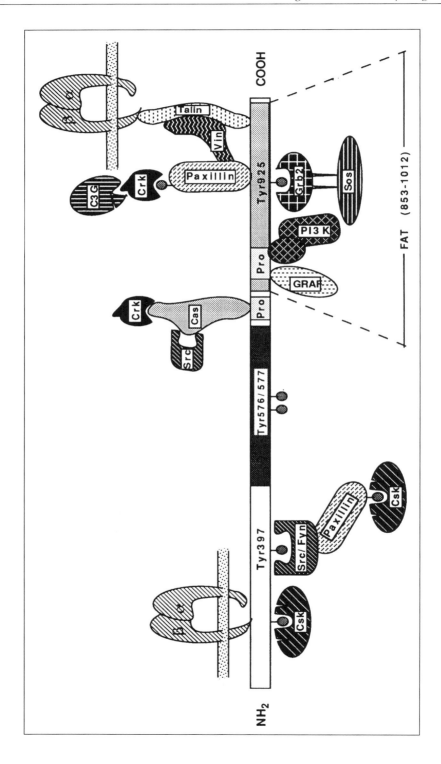

A homolog of the p125Fak, called Fakb, has been detected in human T and B lymphocytes (Kanner et al 1994). The precise structure of Fakb is unknown, but on the basis of binding studies with different anti-peptide sera, it seems to differ from p125Fak in the carboxyterminal domain.

The recently discovered Pyk2, Cakβ and the Raft kinases are highly homologous to the p125Fak (Lev et al 1995; Sasaki et al 1995; Avraham et al 1995). Pyk2 is expressed in brain and kidney (Lev et al 1995) as well as in leukocytes (Tokiwa et al 1996; Gismondi et al submitted). Cakβ is expressed in brain and fibroblasts (Sasaki et al 1995). Raftk kinase has been identified in megakaryocytes, platelets and brain tissues (Li et al 1996). Comparison of the sequences of Pyk2, Cakβ and Raftk indicates that these proteins are indeed the same molecule (Hiregowdara et al 1997).

TYROSINE PHOSPHORYLATION OF FAK FAMILY: REGULATION BY ADHESIVE EVENTS

Earlier works showed that a group of proteins migrating with apparent molecular weight of 100-130 kDa is tyrosine phosphorylated in response to cell adhesion to fibronectin (Guan et al 1991; Kornberg et al 1991) (see also Fig. 2.1). p125Fak is the major component of this complex (Kornberg et al 1992; Burridge et al 1992; Guan and Shalloway, 1992; Hanks et al 1992; Lipfert et al 1992) (Fig. 2.3). Induction of p125Fak tyrosine phosphorylation is an early event, which occurs 30 sec after adhesion and reaches maximum in 15 min (Kornberg et al 1992; Burridge et al 1992; Defilippi et al 1994). Phosphorylation levels then remain high, as long as the cells are attached to the substratum, but drop to minimal levels when cells are brought in suspension (Defilippi et al 1994; Schlaepfer et al 1994).

Several components of the extracellular matrix such as fibronectin, vitronectin, laminin and collagen IV are equally able to trigger p125Fak tyrosine phosphorylation (Kornberg et al 1992; Burridge et al 1992; Defilippi et al 1994; Bozzo et al 1994). p125Fak tyrosine phosphorylation can also be triggered after cell adhesion to dishes coated with anti-integrin monoclonal antibodies or antibody-induced integrin clustering on cells in suspension (Defilippi et al 1994). In platelets, occupancy and clustering of the αIIbβ3 fibrinogen receptor is not sufficient to induce tyrosine phosphorylation of p125Fak; additional stimulation with agonists, such as thrombin, epinephrine or ADP, is required (Haimovich et al 1993; Shattil et al 1994b). The dependence of p125Fak tyrosine phosphorylation on a second stimulus is a specific property of platelets, since ectopic expression of the fibrinogen receptor αIIbβ3 in other cell types shows that integrin clustering is sufficient to trigger p125Fak tyrosine phosphorylation (Pelletier et al 1995; Leong et al 1995; Defilippi et al 1997).

Fig. 2.2. (Opposite) Structure of the p125Fak molecule and its potential binding sites. p125Fak is composed by the central kinase catalytic domain (black box), flanked by large aminoterminal (open box) and carboxyterminal sequences. The aminoterminal domain of p125Fak contains the binding site for the integrin cytoplasmic domain and the tyrosine residue 397, which upon phosphorylation can bind to the SH2 domain of Src kinases. Other tyrosine residues may serve as binding site for Csk. The catalytic domain contains two tyrosine residues, Tyr-576 and Tyr-577, whose phosphorylation regulates the kinase activity of the molecule. The carboxyterminal domain is characterized by two proline-rich regions that can be responsible for binding of SH3 domain-containing proteins, such as Graf-2, p130Cas and the p85 subunit of PI-3 kinase. Tyr-925 when phosphorylated can bind the SH2 domain of Grb-2. In addition the carboxyterminal domain contains binding sites for talin and paxillin as well as the focal adhesion targeting (FAT) sequence, responsible for the localization of p125Fak to the focal adhesion. The possible complexes originating from the interaction of the different molecules with p125Fak are shown. Vin = vinculin.

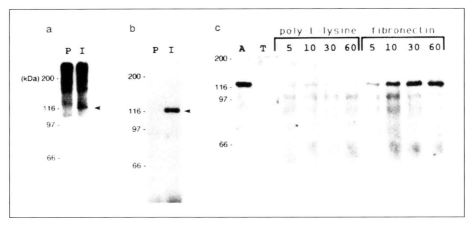

Fig. 2.3. Analysis of p125Fak tyrosine phosphorylation following adhesion to fibronectin in Balb/c 3T3 cells. (a) Immunoprecipitation of p125Fak metabolically labeled with ^{32}P (P: preimmune serum; I: immune serum). (b) Detection of phosphotyrosine in immunoprecipitated p125Fak, using antiphosphotyrosine antibody (P: preimmune serum; I: immune serum). (c) Analysis of p125Fak phosphotyrosine content 5, 10, 30 and 60 min after plating cells onto fibronectin or poly-L-lysine. Lane A: lysate from cells attached and growing under standard culture conditions; lane T: lysate from cells harvested by trypsinization. Arrowheads indicate p125Fak. With permission from Hanks SK et al Proc Natl Acad Sci USA 1992; 89:8487-8491.

β1, β3, β5 and β7 integrin subunits are able to trigger p125Fak tyrosine phosphorylation, while β2 and β4 subunits are not (Akiyama et al 1994; Manie et al 1996). The ability of β1, β3 or β5 integrin subunits to activate p125Fak tyrosine phosphorylation was assessed by mutational analysis of the cytoplasmic domains and by expression of chimeric constructs where the integrin cytoplasmic domains have been fused to an integrin-unrelated extracellular domain. The data showed that the cytoplasmic domains of β subunits are necessary and sufficient to trigger p125Fak activation (Lukashev et al 1994; Akiyama et al 1994) (Fig. 2.4). The structural organization of the β1 cytoplasmic domain is quite restrictive in regulating p125Fak activation. Deletions of few carboxyterminal residues in chicken β1 cytoplasmic domain result in loss of the ability to initiate p125Fak tyrosine phosphorylation (Guan et al 1991). Moreover, integrin isoforms β1B (Altruda et al 1990) and β3B (van Kuppevelt et al 1989), in which only the carboxyterminal region of the cy-

toplasmic domain differs from the sequence of the β1A and β3A, respectively, do not trigger p125Fak tyrosine phosphorylation (Balzac et al 1994; Akiyama et al 1994). In contrast, the muscle specific β1D isoform, which has a cytoplasmic sequence highly homologous to the β1A, maintains this ability (Belkin et al 1996). On the other hand, a β3 deletion variant containing only the membrane proximal 12 amino acids residues elicits the p125Fak phosphorylation response (Leong et al 1995). The same membrane proximal region binds p125Fak (Schaller et al 1995) and is present in β3B, which does not activate p125Fak tyrosine phosphorylation. It can thus be postulated that the highly conserved membrane proximal region is sufficient to bind and activate p125Fak tyrosine phosphorylation, but the distal carboxyterminal sequence in β3B negatively regulates this function.

Phosphorylation of p125Fak and of p41/43FRNK on serine residues has also been reported following adhesion to fibronectin (Hatai et al 1994; Calalb et al 1995; Schaller et al 1993).

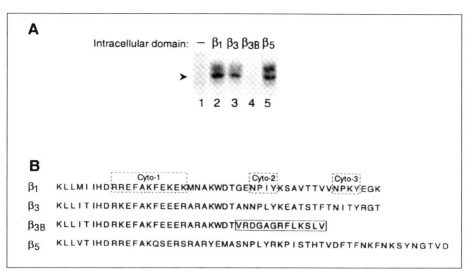

A

Intracellular domain: — β₁ β₃ β₃ᴮ β₅

1 2 3 4 5

B

| | Cyto-1 | Cyto-2 | Cyto-3 |

β₁ KLLMI IHDRREFAKFEKEKMNAKWDTGENPIYKSAVTTVVNPKYEGK

β₃ KLLIT IHDRKEFAKFEEERARAKWDTANNPLYKEATSTFTNITYRGT

β₃ᴮ KLLIT IHDRKEFAKFEEERARAKWDTVRDGAGRFLKSLV

β₅ KLLVT IHDRREFAKQSERSRARYEMASNPLYRKPISTHTVDFTFNKFNKSYNGTVD

Fig. 2.4. p125Fak tyrosine phosphorylation is triggered by specific integrin β intracellular domains in normal human foreskin fibroblasts. (A) Cells transiently expressing the indicated integrin intracellular domains were analyzed for tyrosine phosphorylation. Tyrosine phosphorylation of p125Fak in cells transiently expressing chimeric molecules consisting of the IL-2 receptor extracellular and transmembrane domains either with no integrin intracellular domain or with β1, β3, β3B and β5 intracellular domains are shown in lanes 1-5, respectively. The position of p125Fak is indicated by the arrowhead. The upper phosphorylated band is probably that of p130Cas, which has been observed to be phosphorylated along with p125Fak. (B) Comparison of amino acid sequences of the intracellular domains of β1, β3, β3B and β5 integrins. The dotted box indicates the positions of the cyto-1, cyto-2 and cyto-3 domains. The solid box indicates the portion of the alternatively spliced sequence for the β3B integrin that is nonconserved with respect to the intracellular domains of those integrins capable of participating in transmembrane signaling. With permission from Akiyama SK et al J Biol Chem 1994; 269:15961-15964.

Pyk2 is tyrosine phosphoylated in response to integrin clustering and cell matrix adhesion in natural killer cells, that do not express p125Fak (Gismondi et al. submitted), and in megakaryocytes (Avraham et al 1995).

TYROSINE PHOSPHORYLATION OF Fᴀᴋ FAMILY: REGULATION BY MITOGENS AND GROWTH FACTORS

p125Fak is tyrosine phosphorylated not only in response to integrin-mediated adhesion, but also following v-Src transformation (Fig. 2.5) (Guan and Shalloway, 1992) or cell stimulation with mitogens and growth factors (Rozengurt 1995). Growth factors and mitogens acting either through RTKs, such as platelet-derived growth fac-

tor (Rankin and Rozengurt, 1994) hepatocyte growth factor/scatter factor (Matsumoto et al 1994), and macrophage colony stimulating factor 1 (Kharbanda et al 1995) or through heterotrimeric G protein-coupled receptors, such as bombesin, vasopressin, endothelin, angiotensin II, lysophosphatidic acid, platelet-activating factor and RANTES, are equally able to induce tyrosine phosphorylation of p125Fak (Zachary and Rozengurt 1992; Zachary et al 1993; Sinnett-Smith et al 1993; Polte et al 1994; Seufferlein and Rozengurt, 1994; Soldi et al 1996; Bacon et al 1996). Additional stimuli modulating p125Fak tyrosine phosphorylation include Alzheimer's Aβ peptide (Zhang et al 1994), IL-1 (Arora et al 1995), engagement of FcεRI (Hamawy et al 1993) or T cell receptor (Maguire et al 1995).

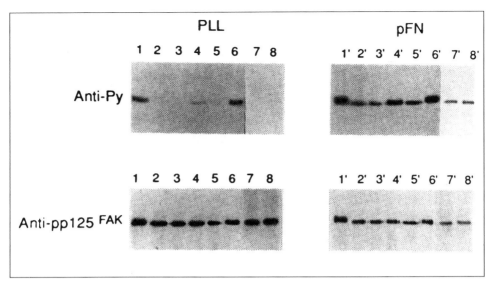

Fig. 2.5. Tyrosine phosphorylation of p125Fak in NIH3T3 cells expressing different Src proteins. p125Fak was immunoprecipitated from various cell lines plated on poly-L-lysine (PLL; left panel) or plasma fibronectin (pFN); right panel). Duplicate samples were electrophoresed on SDS-polyacrylamide gels, transferred to nitrocellulose membranes and probed with either anti-phosphotyrosine antibody (top) or anti p125Fak antibody (bottom). The different forms of Src expressed in NIH3T3 cells are as follows: v-Src (lanes 1, 1'), nonmyristylated c-Src (lanes 2, 2'), kinase defective c-Src (lanes 3, 3'), c-Src (lanes 4, 4'), c-SrcF416 (lanes 5, 5'), and c-SrcF527 (lanes 6, 6'). Also shown is tyrosine phosphorylation of p125Fak in untransfected NIH3T3 cells (lanes 7, 7') and Ras-transformed NIH3T3 cells (lanes 8, 8'). (with permission from Guan JL, Shalloway D. Nature 1992; 358:690-692. Copyright 1992 Macmillan Magazines Limited).

Interestingly, insulin has been reported to cause dephosphorylation of p125Fak (Knight et al 1995; Pillay et al 1995). Tyr-1210 in the insulin receptor cytoplasmic domain affects insulin-induced dephosphorylation of p125Fak but leaves the other insulin-mediated responses intact (Van der Zon, 1996). The insulin-mediated p125Fak dephosphorylation is enhanced by the carboxyterminal Src kinase Csk and involves the PTPase SHPTP2 (Tobe et al 1996; Yamauchi et al 1995).

The convergence on p125Fak of signaling coming from different classes of receptors indicates that this molecule plays an important role in integrating different intracellular signaling pathways. It is still an open question as to whether tyrosine phosphorylation of p125Fak is involved both in the proliferative response to growth factors and in the cytoskeletal reorganization which normally follows growth factor treatment. The PDGF-mediated tyrosine phosphorylation of p125Fak is probably not associated with proliferation, since it occurs at low, not mitogenic doses, of PDGF and is not observed at high mitogenic doses (Rankin and Rozengurt, 1994).

Pyk2 is regulated by stimuli that increase intracellular Ca^{2+} concentration as well as protein kinase C activation, TNFα, ultraviolet radiation and changes in osmolarity (Lev et al 1995; Tokiwa et al 1996). Fakb tyrosine phosphorylation is induced by engagement of TCR/CD3 (Kanner et al 1994).

PATHWAYS LEADING TO P125FAK TYROSINE PHOSPHORYLATION

Intracellular pathways leading to p125Fak tyrosine phosphorylation have not yet been clarified. Four upstream events have been proposed to regulate p125Fak tyrosine phosphorylation: activation of protein kinase C (PKC), mobilization of Ca^{2+}, activation of the small GTPase Rho and integrity of the actin cytoskeleton (see also chapters 3, 4, 6). Involvement of PKC has been demonstrated by showing that specific inhibitors or activators regulate integrin-dependent p125Fak tyrosine phosphorylation (Vuori and Ruoslahti, 1993; Haimovich et al 1996; Defilippi et al 1997). Conflicting results have been reported concerning the role of Ca^{2+}. Ca^{2+} mobilization regulates p125Fak tyrosine phosphorylation following $\alpha IIb\beta 3$ activation and fibrinogen-dependent aggregation in platelets (Shattil et al 1994b) or adhesion to fibrinogen in cells expressing exogenous $\alpha IIb\beta 3$ (Pelletier et al 1992). However, Ca^{2+} is not always required, since integrin-dependent p125Fak tyrosine phosphorylation is not decreased by blocking Ca^{2+} influx in endothelial cells (Schwartz, 1993). p125Fak tyrosine phosphorylation induced by mitogens and growth factors, such as lysophosphatidic acid, bombesin, PDGF, angiotensin II or endothelin, have different requirements for PKC activation and Ca^{2+} mobilization (Zachary et al 1993; Seufferlein and Rozengurt, 1994; Rankin and Rozengurt, 1994; Saville et al 1994; Earp et al 1995; Rozengurt, 1995), indicating that different agonists utilize distinct pathways to induce p125Fak tyrosine phosphorylation.

The small GTP binding protein Rho (see chapter 6) also plays a role in p125Fak tyrosine phosphorylation, as shown by the fact that C3 transferase, a toxin known to inactivate Rho by ADP-ribosylation, decreases the rate of p125Fak tyrosine phosphorylation in response to GTPγS stimulation (Seckl et al 1995). Moreover, introduction of constitutively active Rho into fibroblastic cells increases the level of p125Fak tyrosine phosphorylation (Flinn and Ridley, 1996). Rho has been implicated in bombesin and endothelin-triggered p125FAK tyrosine phosphorylation (Rankin et al 1994). None of these studies address the question whether Rho is involved in integrin-mediated p125Fak tyrosine phosphorylation.

Actin cytoskeleton integrity is a major requirement for p125Fak tyrosine phosphorylation, since depolymerization of actin by cytochalasin D prevents tyrosine phosphorylation of p125Fak in response to both integrins and growth factors (Lipfert et al 1992; Sinnett-Smith et al 1993; Rankin and Rozengurt, 1994). Inhibition of p125Fak tyrosine phosphorylation induced by cytochalasin D treatment is prevented by tyrosine phosphatase inhibitors (Defilippi et al 1995), suggesting that polymerization of actin cytoskeleton can control tyrosine phosphatase activity. Indeed activation of tyrosine phosphatases acting on p125Fak occurs in response to stimuli inducing actin stress fibers and focal adhesions disassembly (Retta et al 1996). Since these stimuli do not affect cell attachment, it is likely that actin cytoskeleton depolymerization per se is responsible for the activation of tyrosine phosphatase activity. These data indicate that inhibition of PTPases by actin cytoskeleton organization may regulate the level of p125Fak tyrosine phosphorylation following adhesion.

P125FAK TYROSINE KINASE ACTIVITY

Tyr-397 is the major site of p125Fak autophosphorylation (Schaller et al 1994; Eide et al 1994) and is phosphorylated following adhesion (Calalb et al 1995) or oncogenic transformation by v-Src (Xing et al 1994). Whether or not the p125Fak tyrosine phosphorylation is essential in the regulation of the kinase activity remains controversial. Mutation of Tyr-397 in phenylalanine (Phe)leads to a 2-fold reduction of kinase activity in vitro (Schaller et al 1994; Eide et al 1995). Other tyrosine residues, Tyr-401, Tyr-576 and Tyr-577, are phosphorylated following adhesion, Tyr-576 and Tyr-577 are also phosphorylated by Src in vitro (Calalb et al 1995). Mutation of Tyr-576 and Tyr-577 to Phe reduces to 50% the kinase activity, while mutations of Tyr-397, Tyr-576 and Tyr-577 drop kinase activity to 20% of the control value. It is thus likely that multiple tyrosine residues are implicated in the regulation of full p125Fak kinase activity (Calalb et al 1995). Tyr-576 and Tyr-577 are localized in the catalytic subdomain VIII (see Fig. 2.2), a region well conserved among different PTK families (reviewed in Hunter 1989) and their phosphorylation might confer an active conformation to this domain for stable interactions with protein substrates.

In vivo and in vitro experiments have demonstrated that paxillin and the p85 subunit of the PI-3 kinase are p125Fak substrates (Turner et al 1993; Chen and Guan, 1994; Bellis et al 1995; Schaller and Parsons, 1995) (see below and chapter 3). Several other molecules, such as tensin, p130Cas, cortactin, SHC and MAPK, whose tyrosine phosphorylation occurs in response to cell-matrix interaction, are also potential p125Fak substrates. This, however, has not been directly demonstrated as PTKs other than p125Fak are also regulated during adhesion.

TYROSINE PHOSPHORYLATION OF Fak FAMILY LEADS TO INTERACTIONS WITH SH2-DOMAIN CONTAINING PROTEINS

Tyrosine phosphorylation creates sites of high-affinity interaction for other molecules containing SH2 or *phospho tyrosine binding* (PTB) domains (Pawson 1995). The SH2 domain of Src forms complexes with Tyr-397 of p125Fak (Schaller et al 1994; Xing et al 1994; Eide et al 1995) (see Fig. 2.2). Fyn and Csk also bind p125Fak via their SH2 domain (Cobb et al 1994; Sabe et al 1994). Binding of Fyn SH2 domain to p125Fak has also been demonstrated in the two hybrid system, using p125Fak cDNA as bait (Polte and Hanks, 1995). The auto-phosphorylation of Tyr-397 is necessary to create a binding site for Src family kinases, since mutation of Tyr-397 to Phe abolishes Src binding (Schaller et al 1994; Cary et al 1996). Src binding to p125Fak activates Src kinase activity by releasing the inhibitory interaction of the Src SH2 domain with the carboxy-terminal Tyr-527 residue (see also Fig. 2.7).

Once activated Src can phosphorylate p125Fak at residues Tyr-407, Tyr-576, Tyr-577 and Tyr-925 (Schlaepfer et al 1994; Calalb et al 1995). The interaction with PTKs of the Src family may also regulate p125Fak kinase activity as indicated by in vivo analysis. Fyn knock-out mice show hypophosphorylation of p125Fak in forebrain protein extracts; the hypophosphorylation correlates with decreased kinase activity, indicating that p125Fak can be regulated by this Src family member (Grant et al 1995). Fig. 2.6 shows binding of c-Src to p125Fak in vivo and in vitro.

Adhesion to fibronectin also promotes in vivo association of p125Fak with the SH2 domain of Grb2 adaptor protein (Schlaepfer et al 1994) (Fig. 2.2). Grb2 (*growth factor receptor-bound protein 2*) is composed exclusively of one SH2 and two SH3 domains. In a classical receptor tyrosine kinase pathway, Grb2 links the activated receptor to Sos, a guanine nucleotide exchange factor, which converts inactive Ras-GDP to active Ras-GTP (see chapter 5). Tyr-925 of p125Fak directly binds to Grb2 SH2 domain in vitro, since mutation to Phe abolishes this interaction (Schlaepfer et al 1994). It has been suggested that binding of Grb2 to p125Fak may lead to activation of the Ras/MAPK pathway. Grb2 binding to p125Fak can also be induced by treatment with macrophage colony stimulating factor (Kharbanda et al 1995). p125Fak may also associate with the SH2 domain of the p85 subunit of PI-3 kinase in cells transformed by polyoma middle t (Bachelot et al 1996).

An SH2-dependent interaction with c-Src, Fyn and Grb2 was also demonstrated for tyrosine phosphorylated Raftk/Pyk2 (Li et al 1996) suggesting that these associations may represent a general mechanism of signal transduction of the Fak family. Fakb, which is tyrosine phosphorylated in T and B cells in response to antigens receptors, associates in vivo with the Zap70 kinase (Kanner et al 1994). The mechanism of Fakb association to Zap70 is not yet known.

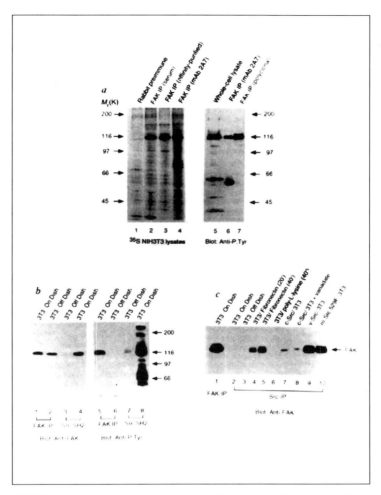

Fig. 2.6. p125Fak association with c-Src in vitro and in vivo. (a) Immunoprecipitation with preimmune serum (lane A), anti p125Fak antiserum (lane 2), affinity-purified anti p125Fak antibodies (lanes 3 and 7) or with anti p125Fak monoclonal antibody 2A7 (lanes 4 and 6) were made from ^{35}S-methionine-labeled NIH3T3 lysates (lanes 1-4) or from unlabeled quiescent NIH3T3 lysates (lanes 6 and 7). NIH3T3 whole cell lysate was also analyzed (lane 5). Immunoprecipitations were visualized by autoradiography (lanes 1-4) or analyzed by antiphosphotyrosine immunoblotting (lanes 5-7). (b) Immunoprecipitation with either anti p125Fak antibodies (lanes 1, 2, 5 and 6) or in vitro binding reactions with GST-Src SH2-domain fusion protein (lanes 3, 4, 7 and 8) were prepared from lysates of NIH3T3 cells grown on plastic (lanes 1, 4, 5, and 8) or detached from plastic (lanes 2, 3, 6, and 7) and analyzed by anti p125Fak (lanes 1-4) followed by antiphosphotyrosine immunoblotting (lanes 5-8). (c) Immunoprecipitations with either anti p125Fak antibodies (lane 1) or anti Src monoclonal antibody 2-17 (lanes 2-10) were prepared from NIH3T3 cells grown on plastic (lanes 1, 2, 7-10), detached from plastic and replated onto fibronectin-coated dishes (lanes 4 and 5) or poly-L-lysine-coated dishes (lane 6). NIH3T3 cells overexpressing chicken Src (lane 7) were incubated with Na₃VO₄ before lysis (lane 8). NIH3T3 cells transformed by v-Src (lane 9) or by activated c-Src (mutation at Tyr-529 to Phe) (lane 10) were used as positive controls for p125Fak coprecipitation with c-Src. (with permission from Schlaepfer DD et al Nature 1994; 372:786-791. Copyright 1994 Macmillan Magazines Limited).

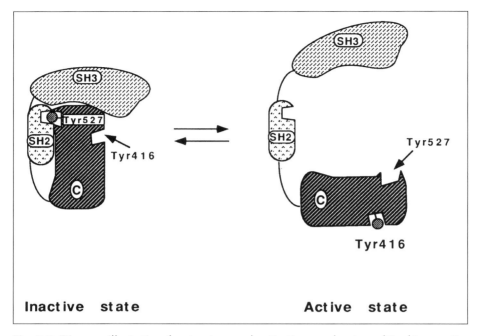

Fig. 2.7. Diagram illustrating the structure and activation mechanism of Src kinases. Src family kinases are characterized by a three domain structure consisting of SH3 and SH2 domains in the aminoterminal half and a catalytic domain (c) at the carboxyiterminal. The phosphorylated carboxyterminal Tyr-527 residue interacts with the amino terminal SH2 domain locking the molecule in a conformation in which both the catalytic (c) and SH2 domain are blocked and the molecule is inactive (left). When Tyr-527 is dephosphorylated the molecule changes its conformation (right) allowing the catalytic domain and the SH2 domain to interact with substrates and other tyrosine phosphorylated proteins respectively. In addition autophosphorylation of Tyr-416 residue is required for full kinase activity.

SH2-INDEPENDENT INTERACTIONS OF FAK FAMILY

In addition to the associations established between tyrosine phosphorylated residues and SH2 domain-containing proteins, p125Fak also binds to several molecules, such as paxillin, talin, p130Cas, Graf-2 and integrins, through structural motifs present in its sequence.

The carboxyterminal domain of p125Fak contains the focal adhesion targeting (FAT) sequence, which resides between amino acids 853 and 1012 (see Fig. 2.2). Addition of the FAT sequence to an unrelated cytosolic protein targets the molecule to focal adhesions, demonstrating that FAT is sufficient to drive intracellular localization of the protein (Hildebrand et al 1993).

It is likely that localization to focal adhesions occurs via binding of FAT sequence to one or several molecules; their identity, however, is still unknown. The carboxyterminal domain binds paxillin and talin (see Fig. 2.2). The paxillin binding site is localized within residues 919-1042 of p125Fak (Tachibana et al 1995); this region partially overlaps the FAT sequence and its relevance to localization in focal adhesions of the p125Fak is debated (Hildebrand et al 1995; Tachibana et al 1995). p125Fak also binds talin in vitro through a 48 amino acid sequence encompassing amino acids 965-1012 of p125Fak. This region is also contained within the FAT sequence, but the talin, binding site is not sufficient per se to direct p125Fak in the focal adhesions (Chen et al 1995). Thus, paxillin, but not talin binding may be relevant to the localization of p125Fak to focal adhesions.

The two proline-rich regions contained in the p125Fak carboxyterminal can bind SH3 domain-containing proteins, such as Graf-2, p85 subunit of PI-3 kinase and p130Cas (see Fig. 2.2). The sequence Pro[875]Lys-Lys-Pro-Pro-Arg of p125Fak binds to the SH3 domain of Graf-2, a novel GTPase-activating protein, whose activity is directed to Cdc42 and RhoA (Hildebrand et al 1996). This proline-rich region also binds in vitro to the SH3 domain of the PI-3 kinase p85 subunit (Guinebault et al 1995). Other seemingly functional proline-rich regions are present on the p125Fak sequence. The amino acid stretch Pro[712]-Pro-Lys-Pro-Ser-Arg has been shown to function as a binding site for the p130Cas SH3 domain and the association between p125Fak and p130Cas has demonstrated in vivo (Polte and Hanks, 1995; Harte et al 1996).

p125Fak can also associate in vitro with the cytoplasmic domain of the β1 integrin subunit (Fig. 2.2). Residues between amino acid 31 and 376 of the aminoterminal domain of p125Fak bind to a peptide reproducing the cytoplasmic domain of the β1 integrin (Schaller et al 1995). Localization to focal adhesion and interaction with integrins appear to be separated functions in the p125Fak molecule since the integrin binding site is located in the aminoterminal domain far from the FAT sequence.

The data discussed above and summarized in Fig. 2.2 indicate that p125Fak contains distinct and partially overlapping sequences able to interact with several molecules, some of which drive localization of p125Fak to the adhesion sites while others are implicated in the downstream signaling. Although most of the interactions have been demonstrated only in in vitro assays, the interaction of Src, paxillin and the p85 subunit of the PI-3 kinase have been confirmed in vivo to be adhesion dependent (Schlaepfer et al 1994; Schaller and Parsons, 1995; Chen and Guan, 1994).

SH2-independent interaction of other Fak family members has not been extensively studied. However, in vivo interaction of paxillin with Pyk2 in natural killer cells has been recently demonstrated (Gismondi et al submitted).

Src KINASES FAMILY

The Src family includes Blk, Fgr, Fyn, Hck, Lck, Lyn, Src, Yes and Yrk kinases (Cooper, 1989). They are anchored to the cytoplasmic side of the plasma membrane via covalent binding of a Gly residue proximal to the aminoterminal of the molecule to a myristic acid. These molecules contain an SH3 and SH2 domains located in the aminoterminal region and a catalytic domain at the carbossiterminal (Erpel and Courtneidge, 1995) (Fig. 2.7). This latter region contains phosphorylation sites involved in regulating kinase activity. In the c-Src prototype, phosphorylation of carboxyterminal Tyr-527 negatively regulates c-Src activity, creating a binding site for the aminoterminal SH2 domain and imposing a conformational change which blocks kinase activation. Phosphorylation of Tyr-527 is regulated by Csk (Inamoto and Soriano 1993). Autophosphorylation of Tyr-416 is required for full kinase activity. The different members of the Src family associate with different receptors and phosphorylate different targets. Lyn, Fyn and Lck kinases, for example, are associated with different sets of receptors in lymphocytes (Weiss and Littman 1994).

INTEGRIN-MEDIATED ACTIVATION OF Src FAMILY KINASES

Adhesion of neutrophils to fibrinogen triggers tyrosine phosphorylation of Fgr and leads to kinase activation (Berton et al 1994). This process can be mimicked by plating cells on dishes coated with antibodies to β2 integrins and does not occur in neutrophils derived from LAD patients that do not express β2 integrins. Activation of Fgr can also be obtained by other stimuli such as TNF, the PKC activator PMA or the chemotactic peptide FMLP (Berton et al 1994). Activation of Fgr in response to these stimuli is mediated through the β2 integrin since it is absent in neutrophils derived from β2-deficient LAD patients. The likely explanation for this phenomenon is that TNF or FMLP trigger inside-out signaling that activates β2 integrin inducing their ability to bind fibrinogen.

Neutrophil spreading triggers translocation of tyrosine phosphorylated proteins and Fgr to detergent-insoluble fraction (Yan et al 1995) suggesting that activation of this kinase may be involved in the organization of the cytoskeleton. However, spreading of neutrophils on anti β2 integrin antibody-coated dishes is not affected in cells deficient in Fgr but is profoundly impaired in cells derived from mice lacking both Fgr and Hck. This suggests that Hck is also involved in integrin-mediated signaling in these cells or that Hck vicariates Fgr in β2 integrin signaling (Lowell et al 1996).

Tyrosine phosphorylation of Fyn and Lck has also been reported upon ligation of α4β1 (VLA-4) in T lymphocytes (Sato et al 1995). Moreover, αV integrin-dependent activation of c-Src has been demonstrated in osteopontin-mediated adhesion of melanoma cells (Chellaiah et al 1996). In these cells, c-Src associates with αVβ3; this mechanism requires the αV integrin cytoplasmic domain, suggesting a direct interaction between integrins and Scr family kinases. It is thus possible that Src kinases participate in integrin signaling either by direct association with integrin cytoplasmic domains or by binding to phosphorylated p125Fak (Schlaepfer et al 1994).

Zap70/Syk FAMILY KINASES

Zap70 and p72Syk are characterized by two SH2 domains, which bind specifically to two closely spaced phosphorylated tyrosine residues inserted in a consensus sequence called ITAM (Immune receptor Tyrosine based Activation Motif) (DeFranco, 1995) present in the cytoplasmic domain of different cell surface receptors or receptor-associated proteins. Binding to the ITAM

sequences leads to activation of Zap70/Syk kinase activity, with subsequent phosphorylation of signaling targets (Weiss and Littman, 1994; DeFranco, 1995). Zap70 is expressed in T and NK cells, whereas p72Syk is expressed preferentially in B cells, myeloid cells, thymocytes and platelets (Weiss and Littman, 1994; Rudd et al 1994).

Zap70 can form stable complexes with p125Fak in T cells following stimulation with the chemochine RANTES (Bacon et al 1996).

p72Syk is rapidly activated following αIIbβ3 or α2β1 crosslinking in platelets (Clark et al 1994; Shattil et al 1994a). The phosphorylation of p72Syk is part of an early wave of tyrosine phosphorylation and precedes the phosphorylation of the p125Fak kinase (Shattil et al 1994a). Regulation of p72Syk tyrosine phosphorylation in platelets is thus distinct from that of p125Fak, since p125Fak phosphorylation requires platelet aggregation. Subsequent to activation, p72Syk kinase activity may be responsible for tyrosine phosphorylation of p140 and p50-68 kDa proteins, which occurs after integrin occupancy (Huang et al 1993; Shattil et al 1994a). Whether the Src kinases are involved in the integrin-mediated p72Syk activation in platelets remains to be determined. β1 integrin-mediated activation of p72Syk has also been reported in monocytes, where it may modulate monocytes functions by leading to activation of NF-κB and increased level of cytokine messages (Lin et al 1995).

In addition to the tyrosine kinases discussed above, integrins regulate tyrosine phosphorylation of several proteins that are likely to be involved in cytoskeletal assembly and intracellular signaling. These include paxillin, p130Cas, cortactin, tensin, PI-3 kinase, Shc and MAPK. Tyrosine phosphorylation of PI-3 kinase, Shc and MAPK are discussed in chapters 3 and 5, while the other proteins are discussed here.

PAXILLIN

Paxillin is a 68 kDa focal adhesion protein initially identified as an highly tyrosine phosphorylated molecule in Rous Sarcoma Virus-transformed cells (Turner et al 1990).

In nontransformed cells, paxillin is rapidly phosphorylated in response to both β1 and β2 integrin stimulation (Burridge et al 1992; Fuortes et al 1994; Rabinovich et al 1995; Leventhal and Feldman, 1996; Gismondi et al submitted). Paxillin is also phosphorylated during embryonic development (Turner et al 1993) and in response to mitogens or growth factors (Zachary et al 1993; Abedi et al 1995; Turner et al 1995).

Paxillin is a substrate of the p125Fak in vitro (Turner et al 1993; Bellis et al 1995) and in vivo (Schaller and Parsons, 1995). Phosphorylation of paxillin occurs at residue Tyr-118 and requires autophosphorylation of p125Fak at Tyr-397 (Bellis et al 1995; Schaller and Parsons, 1995). In vitro paxillin can also be phosphorylated by c-Src or Csk kinases (Schaller and Parsons, 1995). In NK cells, lacking p125Fak, paxillin associates constitutively with Pyk2 and is phosphorylated in response to integrin engagement (Gismondi et al submitted). These data suggest that paxillin can be either directly phosphorylated by Fak family kinases or by PTKs that associate with them. In addition to tyrosine phosphorylation, paxillin may also be serine phosphorylated in response to αVβ5 engagement (De Nichilo and Yamada, 1996).

Immunohistochemical analysis shows that paxillin localizes in the focal adhesions and its sequence predicts the existence of binding sites for several focal adhesion components. A putative SH3-binding proline rich-region, four protein-protein interaction LIM domains (Schmeichel and Beckerle, 1994) and several tyrosine residues as putative binding sites for SH2-containing proteins (Turner and Miller, 1994; Salgia et al 1995) have been identified (Fig. 2.8). Paxillin binds in vitro the membrane proximal region of the β1 integrin (Schaller et al 1995). A 21 amino acid stretch (residues 143-164) in the aminoterminal region of paxillin binds to the carboxyterminal rod domain on the vinculin molecule, thus providing a possible anchorage mechanism of actin filaments to the membrane (Turner et al 1990; Wood et al 1994; Turner and Miller, 1994; Brown et al 1996). The same amino

acid region of paxillin also bind with p125Fak, however, an additional p125Fak binding site is also present encompassing residues 265-313 (Bellis et al 1995; Hildebrand et al 1995; Tachibana et al 1995; Brown et al 1996). These two sites are not crucial for focal adhesion localization of paxillin that depends on the third LIM domain in the carboxyterminal portion of the molecule (Brown et al 1996) (Fig. 2.8). In addition to p125Fak, paxillin also interacts with several cytosolic PTKs including Src (Weng et al 1993), Lyn (Minoguchi et al 1994), Csk (Sabe et al 1994; Bergman et al 1995), the Bcr/Abl oncogene (Salgia et al 1995) as well as the adaptor protein Crk (Birge et al 1993; Schaller and Parsons, 1995) (Fig. 2.8). The Crk SH3 domain can in turn bind to distinct guanine nucleotide exchange factors for Ras, Sos and C3G (Matsuda et al 1994; Tanaka et al 1994) leading to assembly of a transduction machinery that can potentially affect cytoskeleton as well as growth pathways.

P130CAS

p130Cas (Crk-associated substrate) is a 130 kDa protein originally identified as a substrate of v-Src (Kanner et al 1990; Kanner et al 1991) and of v-Crk oncogenes (Sakai et al 1994). In v-Src- and v-Crk-transformed cells tyrosine phosphorylation of p130Cas is constitutive and does not depend on adhesion or cytoskeletal integrity (Sakai et al 1994). In nontransformed cells tyrosine phosphorylation of p130Cas is induced by adhesion to matrix proteins, such as fibronectin, vitronectin, laminin and collagen, in an integrin-dependent manner and is sensitive to cytochalasin D-induced actin cytoskeleton disassembly (Fig. 2.9) (Nojima et al 1995b; Petch et al 1995; Vuori and Ruoslahti, 1995). p130Cas is also tyrosine phosphorylated following interaction of αLβ2 (LFA-1) with ICAM-1 in B-lymphoblastoid cells (Petruzzelli et al 1996). As already described for p125Fak and paxillin, p130Cas is also tyrosine phosphorylated in response to growth factor stimuli (Rozengurt, 1995).

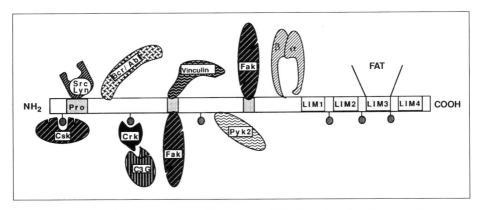

Fig. 2.8. Diagram of paxillin structural features and binding sites. Paxillin contains a putative SH3-binding proline rich-region, four protein-protein interaction LIM domains and several tyrosine residues as putative binding sites for SH2-containing proteins. Potential binding sites for Src kinases, vinculin, p125Fak, Pyk2 and integrins as well as the focal adhesion targeting sequence are shown. Binding to vinculin provides a possible anchorage mechanism of actin filaments to the membrane. Paxillin also binds Csk, the Bcr/Abl oncogene as well as the adaptor protein Crk, which in turn via its SH3 domain can bind to Sos and C3G. The location of the binding sites for Pyk2, Csk, Bcr/Abl and Crk, is hypothetical. (●): phosphorylated tyrosine residue.

c-Src is the kinase responsible for p130Cas phosphorylation induced by adhesion, as shown in c-Src -/- cells, while other PTKs, such as p125Fak, Fyn or Abl, do not seem to be involved (Bockolt and Burridge, 1995; Hamasaki et al 1996; Vuori et al 1996). p130Cas can associate with v-Src and v-Crk (Kanner et al 1991; Sakai et al 1994). Binding to v-Src depends on both SH2 and SH3 domains of Src, whereas v-Crk binds p130Cas through its SH2 domain (Nakamoto et al 1996). p130Cas can also associate with p125Fak in mouse fibroblasts as shown by coprecipitation experiments (Harte et al 1996). This association is likely to involve the interaction of a p130Cas SH3 domain with the amino acid stretch 712-717 of p125Fak as indicated by the two-hybrid system analysis (Polte and Hanks, 1995) (Fig. 2.10). p130Cas together with paxillin may thus be a major protein involved in the formation of the integrin-associated signal transduction complex. Interestingly, Src, Crk and p125Fak, that bind p130Cas, also bind to paxillin, suggesting that p130Cas and paxillin may have partially overlapping functions in integrin-signal transduction.

CORTACTIN

Cortactin is an 80-85 kDa protein that becomes tyrosine phosphorylated in v-Src-transformed cells (Wu et al 1991). Moreover, cortactin is tyrosine phosphorylated in response to integrin-mediated cell adhesion to fibronectin and to immobilized anti-integrin antibodies (Vuori and Ruoslahti, 1995). Cortactin binds to F-actin and the aminoterminal region of the molecule is necessary and sufficient to mediate actin binding. The molecule also contains an SH3 domain in the carboxyterminal region and deletions of this domain have no apparent effect on actin binding activity (Wu and Parsons, 1993).

The role of cortactin in integrin-mediated signaling is unknown. Cortactin is enriched in cortical structures such as membrane ruffles and lamellipodia and does not localize with stress fibers (Wu and Parsons, 1993). Integrin-dependent tyrosine phosphorylation of cortactin may promote its binding to SH2-containing proteins, allowing the protein to act as a docking site for other molecules involved in signaling. Cortactin tyrosine phosphorylation is also

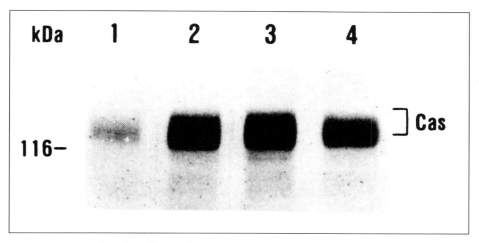

Fig. 2.9. Tyrosine phosphorylation of p130Cas is induced by adhesion to extracellular matrix proteins. p130Cas was immunoprecipitated with a specific antibody from lysates of human skin fibroblasts adhered to poly-L-lysine (lane 1), vitronectin (lane 2), laminin (lane 3) and collagen (lane 4) followed by anti-phosphotyrosine immunoblotting. (with permission from Nojima Y et al J Biol Chem 1995; 270:15398-15402).

stimulated by growth factors, such as PDGF, EGF, colony stimulating factor-1 (Downing and Reynolds, 1992; Maa et al 1992) or thrombin in platelets (Wong et al 1992).

TENSIN

Tensin is a 215 kDa actin nucleating and capping protein containing three actin binding sites and a vinculin binding site. Tensin also contains an SH2 domain, which may interact with tyrosine phosphorylated substrates at focal adhesions (Davis et al 1991; Lo et al 1994). Tyrosine phosphorylation of tensin is dramatically increased in Rous Sarcoma Virus-transformed cells (Davis et al 1991) and is induced by adhesion to matrix proteins in fibroblasts (Bockholt and Burridge, 1993). Increased tensin phosphorylation was also observed during spreading of a colon carcinoma cell line on collagen (Yoshimura et al 1995).

TYROSINE PHOSPHATASES IMPLICATED IN INTEGRIN-MEDIATED SIGNALING

Cellular phosphotyrosine levels are controlled by the coordinate and competing actions of PTKs and PTPases (Brady-Kalnay and Tonks, 1995).

Reduction of cell adhesion by trypsin treatment induces PTPase activity (Maher, 1993). Treatment of cells with PTPase inhibitors leads to increased tyrosine phosphorylation of p125Fak and focal adhesion assembly (Barry and Critchley, 1994; Chrzanowska-Wodnicka and Burridge, 1994; Defilippi et al 1995). PTPase activity, measured as ability to dephosphorylate p125Fak, is inhibited when stress fibers and focal adhesions are assembled (Retta et al 1996). Interestingly the PTPase Lar has been found to localize to focal adhesions, suggesting its possible role in regulating tyrosine phosphorylation events in these structures (Serra-Pages et al 1995). These data strongly suggest a role for PTPases in the regulation of integrin-dependent cytoskeletal organization.

A direct involvement of CD45, a major lymphocyte transmembrane tyrosine phosphatase, in β2 integrin-mediated lymphocyte aggregation has been demonstrated (Arroyo et al 1994). Incubation with antibodies to the extracellular domain of CD45

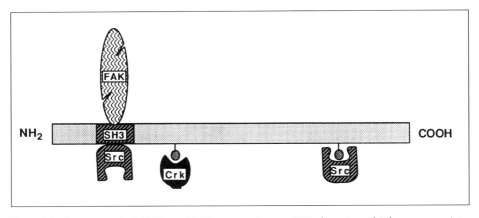

Fig. 2.10. Structure of p130Cas. p130Cas contains an SH3 domain, which can associate with p125Fak. Several tyrosine phosphorylated residues can also associate with v-Src and v-Crk.

inhibits binding of the β2 integrin to the counter receptors ICAM-1 or ICAM-3. CD45 antibody addition also modifies the level of β2 integrin-dependent tyrosine phosphorylation of intracellular proteins (Arroyo et al 1994). These data indicate that CD45 may regulate β2-mediated adhesion, possibly modulating the state of tyrosine phosphorylation of components of the integrin signaling pathway. In platelets, the αIIbβ3 integrin induces cleavage and activation of the cytosolic PTPase PTP1B (Frangioni et al 1993) and PTP1C is tyrosine phosphorylated in response to αIIbβ3-mediated aggregation (Ezumi et al 1995), indicating that integrin αIIbβ3 is involved in the regulation of these PTPases.

ROLE OF PTKS IN INTEGRIN-MEDIATED ADHESION

Tyrosine phosphorylation events in the integrin-mediated signaling are involved in the organization of focal adhesions and actin cytoskeleton as well as in several other processes including activation of the Ras-MAPK pathway, gene expression and apoptosis.

The specific localization of p125Fak and Src in focal adhesions argues for a role of these PTKs in the organization of these adhesive structures. By using drugs to block PTKs or PTPases it was shown that tyrosine phosphorylation of p125Fak correlates with focal adhesion assembly and actin stress fiber organization (Burridge et al 1992; Chrzanowska-Wodnicka and Burridge, 1994; Barry and Critchley, 1994; Defilippi et al 1994 and 1995). Similar correlations have been found with the integrin isoforms β1B, which does not trigger p125Fak tyrosine phosphorylation nor organize focal adhesions (Balzac et al 1994). Further support for the ability of p125Fak to positively influence cell spreading and focal adhesion formation comes from the observation that cells overexpressing the p125Fak isoform p41/43FRNK, which acts as a dominant negative regulator of p125Fak, are delayed in the rate of spreading and focal adhesion organization (Richardson and Parsons, 1996). In apparent contradiction with these results is the finding that embryonal cells lacking p125Fak expression do not have impaired ability to form focal adhesions or induce protein tyrosine phosphorylation in these structures (Ilic et al 1995). A likely explanation for this finding is that other members of the Fak family, such as Pyk2, can substitute or cooperate in the organization of focal adhesions in the absence of p125Fak.

p125Fak also seems to be involved in focal adhesion remodeling during cell motility or cell detachment (Romer et al 1994; Gates et al 1994; Sankar et al 1995). In fact overexpression of p125Fak leads to increased ability of cells to migrate on fibronectin, which is abolished by mutation of Tyr-397, indicating that tyrosine phosphorylation is required for this activity (Cary et al 1996). In addition, microinjection of the carboxy-terminal domain of p125Fak, which mimics the dominant negative FRNK molecule, reduces cell motility and leads to decreased phosphotyrosine content in the focal adhesions (Gilmore and Romer, 1996). Moreover, cells prepared from p125Fak -/- mice exhibit an impaired migratory activity (Ilic et al 1995). These data indicate that localization of p125Fak in focal adhesions and its phosphorylation favor cell migration. Impairment of cell migration in p125Fak -/- cells may explain the defect in mesoderm organization which blocks development of p125Fak knock-out mice (Ilic et al 1995).

p125Fak can contribute to the linkage of actin filaments to the cell membrane through its ability to bind integrins, paxillin and talin. It can be hypothesized that p125Fak binds to the integrin β subunit (Schaller et al 1995) followed by binding of paxillin and talin to p125Fak (Chen et al 1995; Hildebrand et al 1995; Tachibana et al 1995; Bellis et al 1995). Paxillin may thus interacts with the actin-binding molecule vinculin (Turner and Miller 1994) thus creating a linkage between cell membrane and actin cytoskeleton. In a PTK-independent pathway, talin may bind to the β integrin subunit and subsequent binding of vinculin to the complex allows actin recruitment. Fig. 2.11 summarizes two potential pathways leading to focal adhesion assembly.

Src family PTKs have also been implicated in integrin-mediated cell spreading and actin cytoskeleton organization. The first observation comes from studies of v-Src-transformed fibroblasts which exhibit reduced adherence and altered cell morphologies. In these cells, v-Src localizes in residual focal contacts (Rohrschneider, 1980) and in specific adhesive structures called podosomes (Tarone et al 1985). v-Src transformation correlates with disassembly of focal adhesions and with hyper-phosphorylation of integrins and cytoskeletal pro-teins such as paxillin, tensin and p125Fak (Hirst et al 1986; Nigg et al 1986, Turner and Miller, 1994; Lo et al 1994; Guan and Shalloway, 1992). Recently v-Src induced disruption of focal adhesions has been correlated to degradation of the p125Fak kinase, immediately before the onset of cell rounding and detachment induced by v-Src expression (Fincham et al 1995). These data suggest that aberrant activation of Src kinases can profoundly affect the organization of focal adhesions leading to loss of cell spreading. Since Src activity is inhibited by the Csk-dependent phosphorylation of Tyr-527, inhibition of Csk activity might mimic the cytoskeletal reorganization imposed by v-Src. Indeed, fibroblasts derived from Csk knock-out mice show hyperphosphoryl-ation of paxillin, tensin, cortactin and p125Fak and, as a consequence of up regulated Src and Fyn kinase activity, assemble abnormal focal adhesions (Thomas et al 1995).

A number of studies indicate a positive role of the nontransforming c-Src in assembly of adhesive structures. c-Src is normally associated with endosomal membranes and it redistributes to focal adhesions when Tyr-527 is dephosphorylated (Kaplan et al 1994). It has been proposed that dephosphorylation of c-Src Tyr-527 may disrupt SH2-based interactions, changing the conformation of the protein and exposing signals, which contribute to relocalization in focal adhesions. Mutational analysis indicates that the signal responsible for the localization to focal adhesions resides in the first 251 amino acid residues of the aminoterminal domain (Kaplan et al 1994). This region contains both SH3 and SH2 domains, but the SH3 domain is sufficient for localization in focal contacts (Kaplan et al 1994). Neither the SH2 domain nor kinase activity are required for this relocalization. Putative focal adhesion ligands for the c-Src SH3 domain are paxillin and p130Cas.

The localization of c-Src to focal adhesions results in increased detergent insolubility of the molecule (Kaplan et al 1994) consistent with its association to cytoskeletal

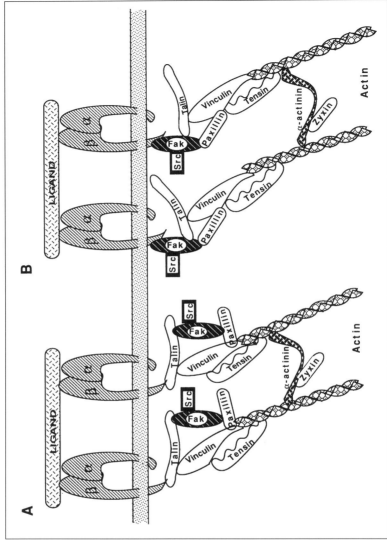

Fig. 2.11. Model of focal adhesion. Integrins in the membrane are connected to the actin filaments via a complex array of bridging molecules. Two hypothetical models of focal adhesion assembly can be proposed on the basis of molecular binding studies discussed in the text. 1) The integrin cytoplasmic domain interacts with talin, which associates with p125Fak and vinculin. The latter two molecules associate with Src, paxillin and tensin, thus establishing a connection with actin filaments (left side). 2) Alternatively, the integrin cytoplasmic domain binds directly to p125Fak, which in turn binds to Src, paxillin and talin. Association of paxillin to p125Fak allows subsequent binding of vinculin, regulating anchorage to actin filaments. (See also Figs. 2.2 and 2.7).

structures. Fibroblasts from c-Src knock-out mice exhibit reduced rate of spreading on fibronectin and the defect can be rescued by the expression of c-Src (Kaplan et al 1995). Mutational analysis indicates that c-Src-induced cell spreading is independent from kinase activity and requires SH2 and SH3 domains. The SH2 domain is necessary for cell spreading, while SH3 domain is required for focal adhesion localization (Kaplan et al 1995). The ability of c-Src SH2 domain to bind p125Fak in response to fibronectin (Cobb et al 1994; Schaller et al 1994; Schlaepfer et al 1994; Xing et al 1994; Eide et al 1995) suggests that p125Fak mediates the effect of c-Src on cell spreading. Once bound to p125Fak, c-Src can phosphorylate different substrates; p130Cas may represent a primary c-Src substrate, since adhesion-induced tyrosine phosphorylation of p130Cas is compromised in mutant cell lines derived from c-Src$^{-/-}$ mice (Bockholt and Burridge, 1995; Vuori et al 1996; Hamasaki et al 1996).

In conclusion a model can be proposed in which c-Src localizes at focal adhesions by an SH3-mediated interaction with paxillin. Once in focal adhesions, by an SH2-mediated interaction with p125Fak, c-Src can induce tyrosine phosphorylation of target molecules, such as p130Cas and p125Fak itself, thus promoting actin skeleton organization and cell spreading.

The opposite effects of v-Src, which reduces adhesion, and c-Src, which enhances the ability of cells to spread, raise unresolved questions on the ability of these two molecules to interact with the regulatory and structural components of focal adhesions. The ability of v-Src to induce degradation of the p125Fak kinase may represent a specific property of transforming Src (Fincham et al 1995). However, further analysis of the different kinase activities for transforming and nontransforming Src molecules and the changes in tyrosine phosphorylation of effector proteins located in focal adhesions may lead to better understanding of the role for these PTKs in the regulation of actin assembly during cell adhesion.

The assembly of cytoskeletal complexes can thus trigger downstream signaling events, leading to control of gene expression, cell proliferation and apoptosis. Integrin-dependent Src binding to p125Fak may trigger phosphorylation of p125Fak Tyr-925, which binds to Grb2/Sos complex (Schlaepfer et al 1994), thus possibly leading to Ras activation. The adaptor protein Crk may also regulate the integrin-mediated activation of Ras. In fact, Crk SH2 domain can associate with tyrosine phosphorylated paxillin (Birge et al 1993; Schaller and Parsons, 1995) or p130Cas (Sakai et al 1994), while the SH3 domain of Crk can interact with C3G, a putative exchange factor for Ras. Sos/Grb2 and C3G/Crk complexes may thus link integrins to the Ras signaling pathways through p125Fak or paxillin and p130Cas. It is interesting to note that dominant negative Ras can block adhesion-dependent activation of MAPK but not p125Fak activation, spreading and stress fibers organization (Clark and Hynes, 1996). These data suggest that Ras is not upstream of p125Fak and integrin-mediated signals regulating actin cytoskeleton may diverge from those regulating MAPK activation at a level upstream of Ras.

Since Ras may be an effector of PI-3 kinase, interaction of p125Fak with PI-3 kinase may represent an alternative way to activate the MAPK pathway (see also chapter 3).

Integrin-induced gene expression requires PTKs activity. PTK inhibitors, genistein and herbimycin A, can block β1 integrin-mediated IL-1β message induction in monocytes (Lin et al 1995). This event seems to require p72Syk rather than p125Fak.

PTK activation can also protect cells from apoptosis (Meredith et al 1994) (see also chapter 8). A direct role of p125Fak in this pathway has been demonstrated in epithelial cells (Frisch et al 1996), although this may not apply to all cell types (Zhang et al 1995b). In this system expression of constitutively active p125Fak does not lead to Erk MAPK activation, suggesting that rescue from apoptosis is not mediated by the Ras/MAPK pathway (Frisch et al 1996). p125Fak is also proteolitically degraded in the early phases of c-myc-induced apoptosis (Crouch et al 1996). p125Fak proteolysis occurs in adherent cells prior to death rather than being a consequence of commitment to dying, suggesting that p125Fak degradation contributes to apoptosis. β1 integrin binding protects cells from apoptosis and also preserves p125Fak from proteolysis (Crouch et al 1996). Likewise, attenuation of p125Fak expression in tumor cells induces apoptosis (Xu et al 1996).

REFERENCES

Abedi H, Daews KE, Zachary I (1995) Differential effects of platelet-derived growth factor BB on p125FAK and paxillin tyrosine phosphorylation and on cell migration in rabbit aortic vascular smooth muscle cells and Swiss 3T3 fibroblasts. J Biol Chem 270:11367-11376.

Akiyama SK, Yamada SS, Yamada KM, LaFlamme SE (1994) Transmembrane signal transduction by integrin cytoplasmic domain expressed in single-subunit chimera. J Biol Chem 269:15961-15964.

Altruda F, Cervella P, Tarone G, Botta C, Balzac F, Stefanuto G, Silengo L (1990) A human integrin β1 subunit with a unique cytoplasmic domain generated by alternative mRNA processing. Gene 95:261-266.

Andrè E, Becker-Andrè M (1993) Expression of an N-terminally truncated form of human focal adhesion kinase in the brain. Biochem Biophys Res Comm 190:140-146.

Arora PD, Ma J, Min W, Cruz T, McCullough CAG (1995) Interleukin-1-induced calcium flux in human fibroblasts is mediated through focal adhesions. J Biol Chem 270:6042-6049.

Arroyo AG, Campanero MR, Sanchez-Mateos P, Zapata JM, Angeles Ursa MA, Del Pozo MA, Sanchez-Madrid F (1994) Induction of tyrosine phosphorylation during ICAM-3 and LFA-1-mediated intercellular adhesion, and its regulation by the CD45 tyrosine phosphatase. J Cell Biol 126:1277-1286.

Avraham S, London R, Fu Y, Ota S, Hiregowdara D, Li J, Jang S, Pasztor LM, White RA, Groopman JE, Avraham H (1995) Identification and characterization of a novel related adhesion focal tyrosine kinase (RAFTK) from megakaryocytes and brain. J Biol Chem 270:27742-27745.

Bachelot C, Rameh L, Parsons T, Cantley LC (1996) Association of phosphatidylinositol 3-kinase, via the SH2 domains of p85, with focal adhesion kinase in polyoma middle t-transformed fibroblasts. Biochim Biophys Acta 1311:45-52.

Bacon KB, Szabo MC, Yssel H, Bolen JB, Schall TJ (1996) RANTES induces tyrosine kinase activity of stable complexes p125Fak and Zap70 in human T cells. J Exp Med 184:873-872.

Balzac F, Retta SF, Albini A, Melchiorri A, Koteliansky VE, Geuna M, Silengo L, Tarone G (1994) Expression of β1B integrin isoform in CHO cells results in a dominant negative effect on cell adhesion and motility. J Cell Biol 127:557-565.

Barry ST, Critchley DR (1994) The RhoA-dependent assembly of focal adhesions in Swiss 3T3 cells is associated with increased tyrosine phosphorylation and the recruitment of both p125FAK and protein kinase C-delta to focal adhesions. J Cell Sci 107:1033-1045.

Belkin AM, Zhidkova NI, Balzac F, Altruda F, Tomatis D, Maier A, Tarone G, Koteliansky VE, Burridge K (1996) β1D integrin displaces the β1A isoform in striated muscles: localization at junctional structures and signaling potential in non-muscle cells. J Cell Biol 132:211-226.

Bellis SL, Miller JT, Turner CE (1995) Characterization of tyrosine phosphorylation of paxillin in vitro by focal adhesion kinase. J Biol Chem 270:17437-17441.

Bergman M, Joukov V, Virtanen I, Alitalo K (1995) Overexpressed Csk tyrosine kinase is localized in focal adhesions, causes reorganization of αvβ5 integrin, and interferes with HeLa cell spreading. Mol Cell Biol 15:711-722.

Berton G, Fumagalli L, Laudanna C, Sorio C (1994) β2 integrin-dependent protein tyrosine phosphorylation and activation of the Fgr protein tyrosine kinase in human neutrophils. J Cell Biol 126:1111-1121.

Birge RB, Fajardo JE, Reichman C, Shoelson SE, Songyang Z, Cantley LC, Hanafusa H

(1993) Identification and characterization of a high-affinity interaction between v-Crk and tyrosine-phosphorylated paxillin in CT10-transformed fibroblasts. Mol Cell Biol 13:4648-4656.

Bockholt SM, Burridge K (1993) Cell spreading on extracellular matrix proteins induces tyrosine phosphorylation of tensin. J Biol Chem 268:14565-14567.

Bockholt SM, Burridge K (1995) An examination of focal adhesion formation and tyrosine phosphorylation in fibroblats isolated from src⁻, fyn⁻ and yes⁻ mice. Cell Adh Comm 3:91-100.

Bozzo C, Defilippi P, Silengo L, Tarone G (1994) Role of tyrosine phosphorylation in matrix-induced neurite outgrowth in human neuroblastoma cells. Exp Cell Res 214:313-322.

Brady-Kalnay SM, Tonks N (1995) Protein tyrosine phosphatases as adhesion receptors. Curr Opin Cell Biol 7:650-657.

Brando C, Shevach EM (1995) Engagement of the vitronectin receptor ($\alpha V\beta 3$) on murine T cells stimulates tyrosine phosphorylation of a 115-kDa protein. J Immunol 154:2005-2011.

Brown MC, Perrotta JA, Turner CE (1996) Identification of LIM3 as the principal determinant of paxillin focal adhesion localization and characterization of a novel motif on paxillin directing vinculin and focal adhesion kinase binding. J Cell Biol 135:1109-1123.

Burridge CA, Turner C, Romer L (1992) Tyrosine phosphorylation of paxillin and pp125FAK accompanies cell adhesion to extracellular matrix: a role in cytoskeletal assembly. J Cell Biol 119:893-904.

Calalb MB, Polte TR, Hanks SK (1995) Tyrosine phosphorylation of focal adhesion kinase at sites in the catalytic domain regulates kinase activity: a role for Src family kinases. Mol Cell Biol 15:954-63.

Cary LA, Chang JF, Guan JL (1996) Stimulation of cells migration by overexpression of focal adhesion kinase and its association with Src and Fyn. J Cell Sci 109:1787-1794.

Chellaiah M, Fitzgerald C, Filardo EJ, Cheresh DA, Hruska KA (1996) Osteopontin activation of c-Src in human melanoma cells requires the cytoplasmic domain of the integrin αv subunit. Endocrinol 137:2432-2440.

Chen HC, Appeddu PA, Parsons JT, Hildebrand JD, Schaller MD, Guan JL (1995) Interaction of focal adhesion kinase with cytoskeletal protein talin. J Biol Chem 270:16995-16999.

Chen HC, Guan JL (1994) Association of focal adhesion kinase with its potential substrate phosphatidylinositol 3-kinase. Proc Natl Acad Sci USA 91:10148-10152.

Choi K, Kennedy M, Keller G (1993) Expression of a gene encoding a unique protein-tyrosine kinase within specific fetal and adult-derived hematopoietic lineages. Proc Natl Acad Sci USA 90:5747-51.

Chrzanowska-Wodnicka M, Burridge K (1994) Tyrosine phosphorylation is involved in reorganization of the actin cytoskeleton in response to serum or LPA stimulation. J Cell Sci 107:3643-3654.

Clark EA, Hynes RO (1996) Ras activation is necessary for integrin-mediated activation of extracellular signal-regulated kinase 2 and cytosolic phospholipase A2 but not for cytoskeletal organization. J Biol Chem 271:14814-14818.

Clark EA, Shattil SJ, Ginsberg MH, Bolen J, Brugge JS (1994) Regulation of the protein tyrosine kinase pp72syk by platelet agonists and the integrin $\alpha IIb\beta 3$: J Biol Chem 269:28859-28864.

Cobb BS, Schaller MD, Leu TH, Parsons JT (1994) Stable association of pp60Src and Fyn with the focal adhesion-associated protein tyrosine kinase pp125FAK. Mol Cell Biol 14:147-155.

Cooper JA (1989) In: Kemp B, Alewood PF, eds. The Src family of protein tyrosine kinases. Boca Raton: CRC Press, 85-113.

Crouch DH, Fincham VJ, Frame MC (1996) Targeted proteolysis of the focal adhesion kinase pp125FAK during c-myc-induced apoptosis is suppressed by integrin signaling. Oncogene 12:2689-2696.

Davis S, Lu ML, Lo SH, Lin S, Butler JA, Druker BJ, Roberts TM, An Q, Chen LB (1991) Presence of an SH2 domain in the actin-binding protein tensin. Science 252:712-715.

De Nichilo MO, Yamada KM (1996) Integrin $\alpha V\beta 5$-dependent serine phosphorylation of paxillin in cultured human macrophages adherent to vitronectin. J Biol Chem 271:11016-11022.

Defilippi, P, Bozzo C, Volpe, G, Romano, G, Venturino, M, Silengo, M, Tarone G (1994) Integrin-mediated signal transduction in human endothelial cells: analysis of tyrosine phosphorylation events. Cell Adh Comm 2:75-86.

Defilippi, P, Retta, F S, Olivo, C, Palmieri, M, Venturino, M, Silengo, L, Tarone G (1995) p125FAK tyrosine phosphorylation and focal adhesion assembly, studies with phosphotyrosine phosphatase inhibitors. Exp Cell Res 221:141-152.

Defilippi, P, Venturino M, Gulino D, Duperrey A, Volpe G, Palmieri M, Boquet P, Fiorentini C, Silengo, L, Tarone G (1997) Dissection of pathways implicated in integrin-mediated actin cytoskeleton assembly: involvement of protein kinase C, RhoGTPase and tyrosine phosphorylation. J Biol Chem in press.

DeFranco AL (1995) Transmembrane signaling by antigen receptors of B and T lymphocytes. Curr Opin Cell Biol 7:163-175.

Downing JR, Reynolds AB (1992) PDGF, CSF-1 and EGF induce tyrosine phosphorylation of p120, a pp60Src transformation associated substrate. Oncogene 6:607-613.

Earp HS, Huckle WR, Dawson TL, Li X, Graves LM, Dy R (1995) Angiotensin II activates at least two tyrosine kinases in rat liver epithelial cells. Separation of the major calcium-regulated tyrosine kinase from p125FAK. J Biol Chem 270:28440-28447.

Eide BL, Turck CW, Escobedo JA (1995) Identification of Tyr-397 as the primary site of tyrosine phosphorylation and pp60Src association in the focal adhesion kinase, pp125FAK. Mol Cell Biol 15:2819-2822.

Erpel T, Courtneidge SA (1995) Src family protein tyrosine kinases and cellular signal transduction pathways. Curr Opin Cell Biol 7:2, 176-182.

Ezumi Y, Takayama H, Okuma M (1995) Differential regulation of protein-tyrosine phosphatases by integrin $\alpha IIb/\beta 3$ through cytoskeletal reorganization and tyrosine phosphorylation in human platelets. J Biol Chem 270:11927-11934.

Ferrel JE, Martin GS (1989) Tyrosine-specific protein phosphorylation is regulated by glycoprotein IIb-IIIa in platelets. Proc Natl Acad Sci USA 86:2234-2238.

Fincham VJ, Wyke JA, Frame MC (1995) v-Src-induced degradation of focal adhesion kinase during morphological transformation of chicken embryo fibroblasts. Oncogene 10:2247-2252.

Flinn HM, Ridley AJ (1996) Rho stimulates tyrosine phosphorylation of focal adhesion kinase, p130 and paxillin. J Cell Sci 109:1133-1141.

Frangioni JV, Oda A, Smith M, Salzman EW, Neel BG (1993) Calpain-catalyzed cleavage and subcellular relocation of protein phosphotyrosine phosphatase 1B (PTP1B) in human platelets. EMBO J 12:4843-4856.

Freedman AS, Rhynhart K, Nojima Y, Svahn J, Eliseo L, Benjamin CD, Morimoto C, Vivier (1993) E Stimulation of protein tyrosine phosphorylation in human B cells after ligation of the $\beta 1$ integrin VLA-4. J Immunol 150:1645-1650.

Frisch AM, Vuori K, Ruoslahti E, Chan-Yui PY (1996) Control of adhesion-dependent cells survival by focal adhesion kinase. J Cell Biol 134:793-799.

Fuortes M, Jin WW, Nathan C (1994) $\beta 2$ integrin-dependent tyrosine phosphorylation of paxillin in human neutrophils treated with tumor necrosis factor. J Cell Biol 127:1477-1483.

Gates RE, King LE Jr, Hanks SK, Nanney LB (1994) Potential role for focal adhesion kinase in migrating and proliferating keratinocytes near epidermal wounds and in culture. Cell Growth Differ 5:891-899.

Gilmore AP, Romer LH (1996) Inhibition of focal adhesion kinase (FAK) signaling in focal adhesions decreases cell motility and proliferation. Mol Biol Cell 7:1209-1224.

Gismondi A, Bisogno L, Palmieri G, Piccoli M, Frati L, Santoni A PYK-2 tyrosine phosphorylation by $\beta 1$ integrin fibronectin receptor crosslinking and association with paxillin in human NK cells. Submitted.

Gismondi A, Milella M, Palmieri G, Piccoli M, Frati L, Santoni A (1995) Stimulation of protein tyrosine phosphorylation by interaction of NK cells with fibronectin via $\alpha 4/\beta 1$ and $\alpha 5/\beta 1$. J Immunol 154:3128-3137.

Golden A, Brugge JS, Shattil SJ (1990) Role of platelet membrane glycoprotein IIb-IIIa in agonist-induced tyrosine phosphorylation of platelet proteins. J Cell Biol 111:3117-3127.

Grant SG, Karl KA, Kiebler MA, Kandel ER (1995) Focal adhesion kinase in the brain, novel subcellular localization and specific regulation by Fyn tyrosine kinase in mutant mice. Genes Dev 9:1909-1921.

Guan JL, Shalloway D (1992) Regulation of focal adhesion-associated protein tyrosine kinase by both cellular adhesion and oncogenic transformation. Nature 358:690-692.

Guan JL, Trevithick JE, Hynes RO (1991) Fibronectin/integrin interaction induces tyrosine phosphorylation of a 120 kDa protein. Cell Regul 2:951-964.

Guinebault C, Payrastre B, Racaud-Sultan C, Mazarguil H, Breton M, Mauco G, Plantavid M, Chap H (1995) Integrin-dependent translocation of phosphoinositide 3-kinase to the cytoskeleton of thrombin activated platelets involves specific interactions of p85 with actin filaments and focal adhesion kinase. J Cell Biol 129:831-842.

Haimovich B, Kaneshiki N, Ping J (1996) Protein Kinase C regulates tyrosine phosphorylation of pp125FAK in platelets adherent to fibrinogen. Blood 87:152-161.

Haimovich B, Lipfert L, Brugge JS, Shattil SJ (1993) Tyrosine phosphorylation and cytoskeletal reorganization in platelets are triggered by interaction of integrin receptors with immobilized ligands. J Biol Chem 268:15868-15877.

Hamasaki K, Mimura T, Morino N, Furuya H, Nakamoto T, Aizawa S, Morimoto C, Yazaki Y, Hirai H, Nojima Y (1996) Src kinase plays an essential role in integrin-mediated tyrosine phosphorylation of Crk-associated substrate p130Cas. Biochem Biophys Res Commun 222:338-343.

Hamawy MM, Mergenhagen SE, Siraganian RP (1993) Cell adherence to fibronectin and the aggregation of the high-affinity immunoglobulin E receptor synergistically regulate tyrosine phosphorylation of 105-115 kDa proteins. J Biol Chem 268:5227-5233.

Hanks SK, Calalb MB, Harper MC, Patel SK (1992) Focal adhesion protein-tyrosine kinase phosphorylated in response to cell attachment to fibronectin. Proc Natl Acad Sci USA 89:8487-8491.

Harte MT, Hildebrand JD, Burnham MR, Bouton AH, Parsons JT (1996) p130Cas, a substrate associated with v-Src and v-Crk, localizes to focal adhesions and binds to focal adhesion kinase. J Biol Chem 271:13649-13655.

Hatai M, Hashi H, Mogi A, Soga H, Yokota J, Yaoi Y (1994) Stimulation of tyrosine- and serine-phosphorylation of focal adhesion kinase in mouse 3T3 cells by fibronectin and fibroblast growth factor. FEBS Lett 350:113-116.

Hildebrand JD, Schaller MD, Parsons JT (1993) Identification of sequences required for the efficient localization of the focal adhesion kinase, pp125FAK, to cellular focal adhesions. J Cell Biol 123:993-1005.

Hildebrand JD, Schaller MD, Parsons JT (1995) Paxillin, a tyrosine phosphorylated focal adhesion-associated protein binds to the carboxyl terminal domain of focal adhesion kinase. Mol Biol Cell 6:637-647.

Hildebrand JD, Taylor JM, Parsons JT (1996) An SH3 domain-containing GTPase-activating protein for Rho and Cdc42 associates with focal adhesion kinase. Mol Cell Biol 16:3169-3178.

Hiregowdara D, Avraham H, Fu Y, London R, Avraham S (1997). Tyrosine phosphorylation of the related adhesion focal tyrosine kinase in megakaryocytes upon stem cell factor and phorpol myristate acetate stimulation and its association with paxillin. J Biol Chem 272:10804-10810.

Hirst R, Horwitz A, Buck C, Rohrschneider L (1986) Phosphorylation of the fibronectin receptor complex in cells transformed by oncogenes that encodes tyrosine kinases. Proc Natl Acad Sci USA 83:6470-6474.

Huang MM, Lipfert L, Cunningham M, Brugge JS, Ginsberg MH, Shattil SJ (1993) Adhesive ligand binding to integrin αIIb β3 stimulates tyrosine phosphorylation of novel protein substrates before phosphorylation of pp125FAK. J Cell Biol 122:473-483.

Hunter T (1989) Protein modifications, phosphorylation on tyrosine residues. Curr Opin Cell Biol 1:1168-1181.

Ihle JN, Kerr IM (1995) Jaks and Stats in signaling by the cytokine receptor superfamily. Trends Genet 11:69-74.

Illc D, Furuta Y, Kanazawa S, Takeda N, Sobue K, Nakatsuji N, Omura S, Fujimoto J, Okada M, Yamamoto T et al (1995) Reduced cell motility and enhanced focal adhesion contact formation in cells from FAK-deficient mice. Nature 377:539-544.

Inamoto A, Soriano P (1993) Disruption of the csk gene, encoding a negative regulator of Src family tyrosine kinases, leads to neural tube defects and embryonic lethality in mice. Cell 73:1117-1124.

Kanner SB, Aruffo A, Chan PY (1994) Lymphocyte antigen receptor activation of a focal adhesion kinase-related tyrosine kinase substrate. Proc Natl Acad Sci USA 91:10484-10487.

Kanner SB, Reynolds AB, Vines RR, Parsons JT (1990) Monoclonal antibodies to individual tyrosine-phosphorylated protein substrates of oncogene-encoded tyrosine kinases. Proc Natl Acad Sci USA 87:3328-3332.

Kanner SB, Reynolds AB, Wang HC, Vines RR, Parsons JT (1991) The SH2 and SH3 domains of pp60Src direct stable association with tyrosine phosphorylated proteins p130 and p110. EMBO J J 10:1689-1698.

Kaplan KB, Bibbins KE, Swedlow JR, Arnaud M, Morgan DO, Varmus HE (1994) Association of the amino/terminal half of c-Src with focal adhesions alters their properties and is regulated by phosphorylation of tyrosine 527. EMBO J 13:4745-4756.

Kaplan KB, Swedlow JR, Morgan DO, Varmus HE (1995) c-Src enhances the spreading of Src-/- fibroblasts on fibronectin by kinase-independent mechanisms. Genes and Develop 9:1505-1517.

Kharbanda S, Saleem A, Yuan Z, Emoto Y, Prasad KV, Kufe D (1995) Stimulation of human monocytes with macrophage colony-stimulating factor induces a Grb2-mediated association of the focal adhesion kinase pp125FAK and dynamin. Proc Natl Acad Sci USA 92:6132-6136.

Knight JB, Yamauchi K, Pessin JE (1995) Divergent insulin and platelet-derived growth factor regulation of focal adhesion kinase (pp125FAK) tyrosine phosphorylation, and rearrangement of actin stress fibers. J Biol Chem 270:10199-10203.

Kornberg L, Earp HS, Parsons JT, Schaller M, Juliano RL (1992) Cell adhesion or integrin clustering increases phosphorylation of a focal adhesion associated tyrosine kinase. J Biol Chem 267:23439-23442.

Kornberg L, Earp HS, Turner CE, Prockop C, Juliano RL (1991) Signal transduction by integrins: increased protein tyrosine phosphorylation caused by integrin clustering. Proc Natl Acad Sci USA 88:8392-8396.

Leong L, Hughes PE, Schwartz MA, Ginsberg MH, Shattil SJ (1995) Integrin signaling: roles for the cytoplasmic tails of αIIb/β3 in the tyrosine phosphorylation of pp125FAK. J Cell Sci 108:3817-3825.

Lev S, Moreno H, Martinez R, Canoll P, Peles E, Musacchio JM, Plowman GD, Rudy B, Schlessinger J (1995) Protein tyrosine kinase PYK2 involved in Ca++-induced regulation of ion channel and MAP kinase functions. Nature 376:737-745.

Leventhal PS, Feldman EL (1996) Tyrosine phosphorylation and enhanced expression of paxillin during neuronal differentiation in vitro. J Biol Chem 271:5957-5960.

Li J, Avraham H, Rogers RA, Raja S, Avraham S (1996) Characterization of RAFTK, a novel focal adhesion kinase, and its integrin-dependent phosphorylation and activation in megakaryocytes. Blood 88:417-428.

Lin TH, Rosales C, Mondal K, Bolen JB, Haskill S, Juliano RL (1995) Integrin-mediated tyrosine phosphorylation and cytokine message induction in monocytic cells. A possible signaling role for the Syk tyrosine kinase. J Biol Chem 270:16189-97.

Lipfert L, Haimovich B, Schaller MD, Cobb BS, Parsons JT, Brugge JS (1992) Integrin-dependent phosphorylation and activation of the protein tyrosine kinase pp125FAK in platelets. J Cell Biol 119:905-12.

Lo SH, Weisberg E, Chen LB (1994) Tensin: a potential link between the cytoskeleton and signal transduction. Bioessays 16:817-823.

Lowell CA, Fumagalli L, Berton G (1996) Deficiency of Src-family kinases p59/61hck and p58c-fgr results in defective adhesion-dependent neutrophil functions. J Cell Biol 133:895-910.

Lukashev ME, Sheppard S, Pytela R (1994) Disruption of integrin function and induction of tyrosine phosphorylation by the autonomously expressed β1 integrin cytoplasmic domain. J Biol Chem 269:18311-18314.

Maa MM, Wilson KK, Moyers JS, Vines RR, Parsons JT (1992) Identification and characterization of a cytoskeleton-associated, epidermal growth factor sensitive pp60Src substrate. Oncogene 7:2429-2438.

Maguire JE, Danahey KM, Burkly LC, van Seventer G (1995) T cell receptor- and β1 integrin-mediated signals synergize to induce tyrosine phosphorylation of focal adhesion kinase (pp125FAK) in human T cells. J Exp Med 182:2079-2090.

Maher P (1985) Phosphotyrosine-containing proteins are concentrated in focal adhesions and intercellular junctions in normal cells. Proc Natl Acad Sci USA 82:6576-6580.

Maher P (1993) Activation of phosphotyrosine phosphatase activity by reduction of cell-substrate adhesion. Proc Natl Acad Sci USA 90:11177-11181.

Manie SN, Astier A, Wang D, Phifer JS, Chen J, Lazarovits AI, Morimoto C, Freedman AS (1996) Stimulation of tyrosine phosphorylation after ligation of β7 and β1 integrins on human B cells. Blood 87:1855-1861.

Matsuda M, Hashimoto Y, Muroya K, Hasegawa H, Kruata T et al (1994) Crk protein binds to two guanine-nucleotide-releasing proteins for the ras family and modulates nerve growth factor-induced activation of ras in PC12 cells. Mol Cell Biol 14:5495-5500.

Matsumoto K, Nakamura T, Kramer RH (1994) Hepatocyte growth factor/scatter factor induces tyrosine phosphorylation of focal adhesion kinase (p125FAK) and promotes migration and invasion by oral squamous carcinoma cells. J Biol Chem, 269:31807-31813.

Meredith JE, Fazeli B, Schwartz MA (1993) The extracellular matrix as a cells survival factor. Mol Biol Cell 4:953-961

Minoguchi k, Kihara H, Nishikata H, Hamawy MM, Siraganian RP (1994) Src family tyrosine kinase Lyn binds several proteins including paxillin in rat basophilic leukemia cells. Mol Immunol 31:519-529.

Nakamoto T, Sakai R, Ozawa K, Yazaki Y, Hirai H (1996) Direct binding of C-terminal region of p130Cas to SH2 and SH3 domains of Src kinase. J Biol Chem 271:8959-8965.

Nakamura S, Yakamura H (1989) Thrombin and collagen induce rapid phosphorylation of a common set of cellular proteins on tyrosine in human platelets. J Biol Chem 264:7089-7091.

Nigg EA, Sefton BM, Singer SJ, Vogt PK (1986) Cytoskeletal organization, vinculin phosphorylation and fibronectin expression in transformed fibroblasts with different cell morphologies. Virology 151:50-65.

Nojima Y, Noritsugo M, Mimura T, Hamasaki K, Furuya H, Sakai R, Sato T, Tachibana K, Morimoto C, Yazaki Y, Hirai H (1995b) Integrin-mediated cell adhesion promotes tyrosine phosphorylation of p130Cas, a Src homology 3-containing molecule having multiple Src homology 2-binding motifs. J Biol Chem 270:15398-15402.

Nojima Y, Rothstein DM, Sugita K, Schlossman SF, Morimoto C (1992) Ligation of VLA-4 on T cells stimulates tyrosine phosphorylation of a 105 kDa protein. J Exp Med 175:1045-1053.

Nojima Y, Tachibana K, Sato T, Schlossman SF, Morimoto C (1995a) Focal adhesion kinase is tyrosine phosphorylated after engagement of α4/β1 and α5/β1 integrins on human T-lymphoblastic cells. Cell Immunol 161:8-13.

Pawson T (1995) Protein modules and signaling networks. Nature 373:573-580.

Pelletier AJ, Bodary SC, Levinson AD (1992) Signal transduction by the platelet integrin αIIb/β3: induction of calcium oscillations required for protein-tyrosine phosphorylation and ligand-induced spreading of stably transfected cells. Mol Biol Cell 3:989-998.

Pelletier AJ, Kunicki T, Ruggeri ZM, Quaranta V (1995) The activation state of the integrin αIIbβ3 affects outside-in signals leading to cell spreading and focal adhesion kinase phosphorylation. J Biol Chem 270:18133-18140.

Petch LA, Bockholt SM, Bouton A, Parsons JT, Burridge K (1995) Adhesion-induced tyrosine phosphorylation of the p130 Src substrate. J Cell Sci 108:1371-1379.

Petruzzelli L, Takami M, Herrera R (1996) Adhesion through the interaction of lymphocyte function-associated antigen-1 with intracellular adhesion molecule-1 induces tyrosine phosphorylation of p130Cas and its association with c-CrkII. J Biol Chem 271:7796-7801.

Pillay TS, Sasaoka T, Olefsky JM (1995) Insulin stimulates the tyrosine dephosphorylation of pp125 focal adhesion kinase. J Biol Chem 270:991-994.

Polte TR, Hanks SK (1995) Interaction between focal adhesion kinase and Crk-associated tyrosine kinase substrate p130Cas. Proc Natl Acad Sci USA 92:10678-10682.

Polte TR, Naftilan AJ, Hanks SK (1994) Focal adhesion kinase is abundant in developing blood vessels and its elevation in its phosphotyrosine content in vascular smooth muscle cells is a rapid response to angiotensin II. J Cell Biochem 55:106-119.

Rabinovich H, Lin WC, Manciulea M, Herbeman RB, Whiteside TL (1995) Induction of protein tyrosine phosphorylation in human natural killer cells by triggering via $\alpha4\beta1$ or $\alpha5\beta1$ integrins. Blood 85:1858-1864.

Rankin S, Morii N, Narumiya S, Rozengurt E (1994) Botulinum C3 exoenzyme blocks the tyrosine phosphorylation of p125FAK and paxillin induced by bombesin and endothelin. FEBS Let 354:315-319.

Rankin S, Rozengurt E (1994) Platelet-derived growth factor modulation of focal adhesion kinase (p125FAK) and paxillin tyrosine phosphorylation in Swiss 3T3 cells. J Biol Chem 269:704-710.

Retta SF, Barry ST, Critchley DR, Defilippi P, Silengo L, Tarone G (1996) Focal adhesion and stress fiber formation is regulated by tyrosine phosphatase activity. Exp Cell Res 229:307-317.

Richardson A, Parsons JT (1996) A mechanism for regulation of the adhesion-associated protein tyrosine kinase pp125FAK. Nature 380:538-540.

Rohrschneider LR (1980) Adhesion plaques of Rous sarcoma virus-transformed cells contain the Src gene product. Proc Natl Acad Sci USA 77:3514-3518.

Romer LH, McLean N, Turner CE, Burridge K (1994) Tyrosine kinase activity, cytoskeletal organization and motility in human vascular endothelial cells. Mol Biol Cell 6:349-361.

Rozengurt E (1995) Convergent signaling in the action of integrins, neuropeptides, growth factors and oncogenes. Cancer Surv 24:81-96.

Rudd CE, Janssen O, Cai Y, da Silva AJ, Raab M, Prasad KSV (1994) Two-step TCR /CD3-CD4 and CD28 signaling in T-cells: SH2/SH3 domains, protein-tyrosine and lipid kinases. Immunol Today 15:225-234.

Sabe H, Hata A, Okada M, Nakagawa H, Hanafusa H (1994) Analysis of the binding of the Src homology 2 domain of Csk to tyrosine-phosphorylated proteins in the suppression and mitotic activation of c-Src. Proc Natl Acad Sci USA 91:3984-3988.

Sakai R, Iwamatsu A, Hirano N, Ogawa S, Tanaka T, Mano H, Yazaki Y, Hirai H (1994) A novel signaling molecule, p130, forms stable complexes in vivo with v-crk and c-Src in a tyrosine phosphorylation-dependent manner. EMBO J 13:3748-3756.

Salgia R, Li JL, Lo SH, Brunkhorst B, Kansas GS, Sobhany ES Sun Y, Pisick E, Hallek M, Ernst T et al (1995) Molecular cloning of human paxillin, a focal adhesion protein phosphorylated by P210BCR/ABL. J Biol Chem 270:5039-5047.

Sanchez-Mateos P, Campanero MR, Balboa MA, Sanchez-Madrid F (1993) Co-clustering of $\beta1$ integrins, cytoskeletal protein, and tyrosine-phosphorylated substrates during integrin-mediated leukocyte aggregation. J Immunol 151:3817-3828.

Sankar S, Mahooti-Brooks N, Hu G, Madri JA (1995) Modulation of cell spreading and migration by pp125FAK phosphorylation. Am J Pathol 147:601-608.

Sasaki H, Nagura K, Ishino M, Tobioka H, Kotani K, Sasaki T (1995) Cloning and characterization of cell adhesion kinase β, a novel protein-tyrosine kinase of the focal adhesion kinase subfamily. J Biol Chem 270:21206-21219.

Sato T, Tachibana K, Nojima Y, D'Avirro N, Morimoto C (1995) Role of the VLA-4 molecule in T cells costimulation. J Immunol 155:2938-2947.

Saville MK, Graham A, Malarkey K, Paterson A, Gould GW, Plevin R (1995) Regulation of endothelin-1- and lysophosphatidic acid-stimulated tyrosine phosphorylation of focal adhesion kinase (pp125fak) in Rat-1 fibroblasts. Biochem J 301:407-414.

Schaller MD, Borgman CA, Cobb BS, Vines RR, Reynolds AB, Parsons JT (1992) pp125FAK a structurally distinctive protein-tyrosine kinase associated with focal adhesions. Proc Natl Acad Sci USA 89:5192-5196.

Schaller MD, Borgman CA, Parsons JT (1993) Autonomous expression of a non-catalytic domain of the focal adhesion-associated protein tyrosine kinase pp125FAK. Mol Cell Biol 13:785-791.

Schaller MD, Hildebrand JD, Shannon JD, Fox JW, Vines RR, Parsons JT (1994) Autophosphorylation of the focal adhesion kinase, pp125FAK, directs SH2-dependent binding of pp60Src. Mol Cell Biol 14:1680-1688.

Schaller MD, Otey CA, Hildebrand JD, Parsons JT (1995) Focal adhesion kinase and paxillin bind to peptides mimicking β integrin cytoplasmic domains. J Cell Biol 130:1181-1187.

Schaller MD, Parsons JT (1995) pp125FAK-dependent tyrosine phosphorylation of paxillin creates a high-affinity binding site for Crk. Mol Cell Biol 15:2635-2645.

Schlaepfer DD, Hanks SK, Hunter T, Van der Geer P (1994) Integrin-mediated signal transduction linked to Ras pathway by GRB2 binding to focal adhesion kinase. Nature 372:786-791.

Schmeichel KL, Beckerle MC (1994) The LIM domain is a modular protein-binding interface. Cell 79:211-219.

Schwartz MA (1993) Spreading of human endothelial cells on fibronectin or vitronectin triggers elevation of intracellular free calcium. J Cell Biol 120:1003-1010.

Scott G, Liang H (1995) pp125FAK in human melanocytes and melanoma: expression and phosphorylation. Exp Cell Res 219:197-203.

Seckl MJ, Morii N, Narumiya S, Rozengurt E (1995) Guanosine 5'-3-O-(thio)triphosphate stimulates tyrosine phosphorylation of p125FAK and paxillin in permeabilized

Swiss 3T3 cells. Role of p21rho. J Biol Chem 270:6984-6990.

Serra-Pages C, Kedersha NL, Fazikas L, Medley Q, Debant A, Streuli M (1995) The LAR transmembrane protein tyrosine phosphatase and a coiled-coil LAR-interacting protein colocalizes at focal adhesions. EMBO J 14:2827-2838.

Seufferlein T, Rozengurt E (1994) Lysophosphatididc acid stimulates tyrosine phosphorylation of focal adhesion kinase, paxillin and p130. J Biol Chem 269:9345-9351.

Shattil SJ, Ginsberg MH, Brugge JS (1994a) Adhesive signaling in platelets. Curr Opin Cell Biol 6:695-704.

Shattil SJ, Haimovich B, Cunningham M, Lipfert L, Parsons JT, Ginsberg MH, Brugge JS (1994b) Tyrosine phosphorylation of pp125FAK in platelets requires coordinated signaling through integrin and agonist receptor. J Biol Chem 269:14738-14745.

Sinnett-Smith J, Zachary I, Valverde AM, Rozengurt E (1993) Bombesin stimulation of p125 focal adhesion kinase tyrosine phosphorylation. J Biol Chem 268:14261-14268.

Soldi R, Sanavio F, Aglietta M, Primo L, Defilippi P, Marchisio PC, Bussolino F (1996) Platelet-activating factor (PAF) induces the early tyrosine phosphorylation of focal adhesion kinase (p125FAK) in human endothelial cells. Oncogene 13:515-525.

Tachibana K, Sato T, D'Avirro N, Morimoto C (1995) Direct association of pp125FAK with paxillin, the focal adhesion-targeting mechanism of pp125FAK. J Exp Med 182:1089-1099.

Tanaka S, Morishita T, Hashimoto Y, Hattori S, Nalamura S et al (1994) C3G, a guanine nucleotide-releasing protein expressed ubiquitously, binds to the Src homology 3 domains of CRK and GRB2/ASH proteins. Proc Natl Acad Sci USA 91:344-347.

Tarone G, Cirillo D, Giancotti FG, Comoglio PM, Marchisio PC (1985) Rous sarcoma virus-transformed fibroblasts adhere primarily at discrete protrusions of the ventral membrane called podosomes. Exp Cell Res 159:141-157.

Thomas SM, Soriano P, Inamoto A (1995) Specific and redundant roles of Src and Fyn in organizing the cytoskeleton. Nature 376:267-271.

Tobe K, Sabe H, Yamamoto T, Yamauchi T, Asai S, Kaburagi Y, Tamemoto H, Ueki K, Kimura H, Akanuma Y, Yazaki Y, Hanafusa H, Kadowaki T (1996) Csk enhances insulin-stimulated dephosphorylation of focal adhesion proteins. Mol Cell Biol 16:4765-4772.

Tokiwa G, Dikic I, Lev S, Schlessinger J (1996) Activation of Pyk2 by stress signals and coupling with JNK signaling pathway. Science 273:792-794.

Turner CE, Glenney JR, Burridge K (1990) Paxillin: a new vinculin-binding protein present in focal adhesions. J Cell Biol 111:1059-1068.

Turner CE, Miller JT (1994) Primary sequence of paxillin contains putative SH2 and SH3 domain binding motifs and multiple LIM domains: identification of a vinculin and pp125Fak-binding region. J Cell Sci 107:1583-1591.

Turner CE, Pietras KN, Taylor DS, Molloy CJ (1995) Angiotensin II stimulation of rapid paxillin tyrosine phosphorylation correlates with the formation of focal adhesions in rat aortic smooth muscle cells. J Cell Sci 108:333-342.

Turner CE, Schaller MD, Parsons JT (1993) Tyrosine phosphorylation of the focal adhesion kinase pp125FAK during development: relation to paxillin. J Cell Sci 105:637-645.

Van der Geer P, Hunter T, Lindberg RA (1994) Receptor protein-tyrosine kinases and their signal transduction pathways. Annu Rev Cell Biol 10:251-337.

Van der Zon GC, Ouwens DM, Dorrestijn J, Maassen JA (1996) Replacement of the conserved tyrosine 1210 by phenylalanine in the insulin receptor affects insulin-induced dephosphorylation of focal adhesion kinase but leaves other responses intact. Biochemistry 35:10377-10382.

van Kuppevelt TH, Languino LR, Gailit JO, Suzuki S, Ruoslahti E (1989) An alternative cytoplasmic domain of the integrin β3 subunit. Proc Natl Acad Sci USA 86:5415-5418.

Vuori K, Hirai H, Aizawa S, Ruoslahti E (1996) Induction of p130Cas signaling complex formation upon integrin-mediated cell adhesion: a role for Src family kinases. Mol Cell Biol 16:2606-2613.

Vuori K, Ruoslahti E (1993) Activation of protein kinase C precedes α5/β1 integrin-mediated cell spreading. J Biol Chem 268:21459-21462.

Vuori K, Ruoslahti E (1995) Tyrosine phosphorylation of p130Cas and cortactin accompanies integrin-mediated cell adhesion to extracellular matrix. J Biol Chem 270:22259-22262.

Weiss A, Littman DR (1994) Signal transduction by lymphocyte antigen receptors. Cell 76:263-274.

Weng Z, Taylor JA, Turner CE, Brugge JS, Seidel-Dugan C (1993) Detection of Src homology 3-binding proteins, including paxillin, in normal and v-Src-transformed Balb/c 3T3 cells. J Biol Chem 268:14956-63.

Whitney GS, Chan PY, Blake J, Cosand WL, Neubauer MG et al (1993) Human T and B lymphocytes express a structurally conserved focal adhesion kinase, pp125FAK. DNA Cell Biol 10912:823-830.

Wong S, Reynolds AB, Papkoff J (1992) Platelet activation leads to increased c-Src kinase activity and association of c-Src with an 85 kDa tyrosine phosphoprotein. Oncogene 7:2407-2415

Wood CK, Turner CE, Jackson P, Critchley DR (1994) Characterisation of the paxillin-binding site and the C-terminal focal adhesion targeting sequence in vinculin. J Cell Sci 107:709-717.

Wu H, Parsons JT (1993) Cortactin, an 80/85 kDa pp60Src substrate, is a filamentous actin binding protein enriched in the cell cortex. J Cell Biol 120:1417-1426.

Wu H, Reynolds AB, Kanner SB, Vines RR, Parsons JT (1991) Identification of a novel cytoskeleton-associated pp60Src substrate. Mol Cell Biol 11:5113-5124.

Xing Z, Chen HC, Nowlen JK, Taylor SJ, Shalloway D, Guan JL (1994) Direct interaction of v-Src with the focal adhesion kinase mediated by the Src SH2 domain. Mol Biol Cell 5:413-421.

Xu LH, Owens LV, Sturge GC, Yang X, Liu ET, Craven RJ, Cance WG (1996) Attenuation of the expression of the focal adhesion kinase induces apoptosis in tumor cells. Cell Growth Differ 7:413-418.

Yamauchi K, Milarski KL, Saltiel AR, Pessin JE (1995) Protein-tyrosine-phosphatase SHPTP2 is a required positive effector for insulin downstream signaling. Proc Natl Acad Sci USA 92:664-668.

Yan SR, Fumagalli L, Berton G (1995) Tumor necrosis factor triggers redistribution to a Triton X-100-insoluble, cytoskeletal fraction of β2 integrins, NADPH oxidase components, tyrosine phosphorylated proteins, and the protein tyrosine kinase p58fgr in human neutrophils adherent to fibrinogen. J Leuk Biol 58:595-602.

Yoshimura M, Nishikawa A, Nishiura T, Ihara Y, Kanayama Y, Matsuzawa Y, Taniguchi N (1995) Cell spreading in Colo 201 by staurosporin is α3/β1 integrin-mediated with tyrosine phosphorylation of Src and tensin J Biol Chem 270:2298-2304.

Zachary I, Rozengurt E (1992) Focal adhesion kinase (p125FAK): a point of convergence in the action of neuropeptides, integrins, and oncogenes. Cell 71:891-894.

Zachary I, Sinnett-Smith J, Turner CE, Rozengurt E (1993) Bombesin, vasopressin and endothelin rapidly stimulates tyrosine phosphorylation of the focal adhesion-associated protein paxillin in Swiss 3T3 cells. J Biol Chem 268:22060-22065.

Zhang C, Lambert MP, Bunch C, Barber K, Wade WS, Krafft GA, Klein WL (1994) Focal adhesion kinase expressed by nerve cell lines shows increased tyrosine phosphorylation in response to Alzheimer's A β peptide. J Biol Chem 269:25247-25250.

Zhang X, Wright CV, Hanks SK (1995a) Cloning of a Xenopus laevis cDNA encoding focal adhesion kinase (FAK) and expression during early development. Gene 160:219-222.

Zhang Z, Vuori K, Reed JC, Ruoslahti E (1995b) The α5β1 integrin supports survival of cells on fibronectin and up-regulates Bcl-2 expression. Proc Natl Acad Sci USA 92:6161-6165.

Ziemiecki A, Harpur AG, Wilks AF (1994) JAK protein tyrosine kinases: their role in cytokine signaling. Trends in Cell Biol 4:207-212.

CHAPTER 3

Lipid Kinase, Phospholipase and PKC Activation

PHOSPHOINOSITIDE (PI) METABOLISM

In recent years phosphoinositide metabolism has been implicated as a key component of signal transduction cascades initiated by a variety of receptors (Divecha and Irvine, 1995). Two major pathways have been described to date: the canonical PI turnover and the 3-phosphoinositide pathway (Fig. 3.1). In the PI turnover pathway, phosphatidylinositol (PtdIns) is sequentially phosphorylated by PtdIns-4-kinase and Ptd-5-kinase to generate phosphatidylinositol 4,5-biphosphate (PtdIns(4,5)P$_2$), the major substrate of PtdIns-specific phospholipase (PL) C. The rapid activation of PtdIns-specific PLC in response to ligand-receptor interaction results in hydrolysis of PtdIns(4,5)P$_2$ by PLC and generation of the two second messengers, inositol 1,4,5-trisphosphate (Ins(1,4,5)P$_3$) and diacylglycerol (DAG) (see also paragraph "phospholipases"): Ins(1,4,5)P$_3$ promotes release of Ca^{2+} from intracellular stores and DAG activates protein kinase C (PKC).

The 3-phosphoinositide pathway involves PTK-mediated recruitment and activation of PtdIns-3 kinase which phosphorylates PtdIns, PtdIns(4)P and PtdIns(4,5)P$_2$ at the D-3 position of the inositol ring, resulting in the production of PtdIns(3)P, PtdIns(3,4)P$_2$ and PtdIns(3,4,5)P$_3$, respectively. The 3-phosphoinositides are not substrates of any known PLC, or components of the canonical phosphoinositide turnover pathway, but they have been described as second messengers in many cellular responses including mitogenesis, apoptosis, cytoskeletal rearrangements and vesicular trafficking.

Thus, PtdIns(4,5)P$_2$ occupies a central role in the phosphoinositide signal transduction cascade serving as substrate for receptor-activated PLC or PtdIns-3-kinase (Lee and Rhee, 1995) (Fig. 3.2). PtdIns(4,5)P$_2$ itself can also regulate the function of cytoskeletal proteins including profilin, gelsolin, α-actinin or of a number of enzymes, such as PKC, casein kinase I and PLD. Moreover, recent evidence indicates that PtdIns(4,5)P$_2$ binds to pleckstrin homology domains (PH) which are present in many signaling molecules (Harlan et al 1994).

PHOSPHOINOSITIDE KINASES

Phosphoinositide kinases involved in synthesis of polyphosphoinositides include PtdIns-4, PtdIns-5 and PtdIns-3 kinases.

Signal Transduction by Integrins, by Paola Defilippi, Angela Gismondi, Angela Santoni and Guido Tarone. © 1997 Landes Bioscience.

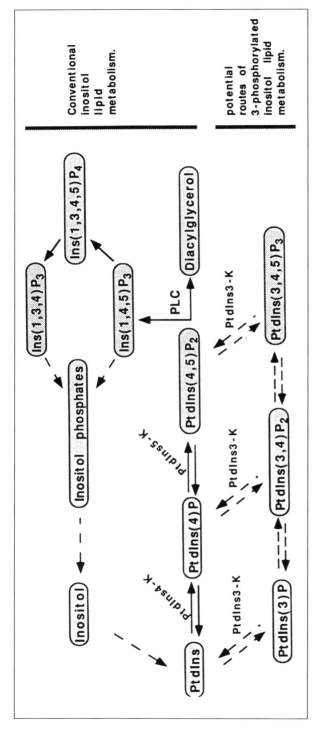

Fig. 3.1. Phosphoinositide metabolism: canonical PI turnover and 3-phosphoinositide pathway. In the PI turnover pathway, phosphatidylinositol (PtdIns) is sequentially phosphorylated by PtdIns-4-kinase and PtdIns-5-kinase to generate phosphatidylinositol 4,5-biphosphate (PtdIns(4,5)P_2), the major substrate of PtdIns-specific phospholipase C (PLC). The activation of PtdIns-specific PLC in response to ligand-receptor interaction results in hydrolysis of PtdIns(4,5)P_2 by PLC and generation of the two second messengers :inositol 1,4,5-triphosphate (Ins(1,4,5)P_3) and diacylglycerol (DAG) Ins(1,4,5)P_3 promotes release of Ca^{2+} from intracellular stores and DAG activates protein kinase C (PKC). The 3-phosphonositide pathway involves activation of PtdIns-3-kinase which phosphorylated PtdIns, PtdIns(4)P and PtdIns(4,5)P_2 at D-3 position of the inositol ring, resulting in the production of PtdIns(3)P, PtdIns(3,4)P_2 and PtdIns(3,4,5)P_3, respectively.

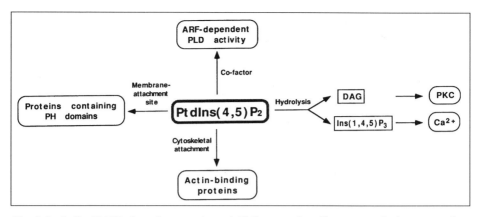

Fig. 3.2. PtdIns(4,5)P$_2$ functions: activated PLC exerts its effect not only by generating messenger molecules, DAG and Ins(1,4,5)P$_3$, but also by reducing the levels of PtdIns(4,5)P$_2$, which serves as a cofactor of ARF (ADP-ribosylation factor)-dependent phopholipase D (PLD), as a membrane attachment site for proteins containing pleckstrin homology (PH) domains and as a regulator of actin-binding proteins.

Multiple cellular forms of PtdIns-4 and PtdIns-5 kinases have been isolated from mammalian sources as well as from yeast (Carpenter and Cantley, 1990). In regard to PtdIns-5 kinase, two isoforms using PtdIns(4)P as substrate, denoted types I and II, have been distinguished by lack of immunocrossreactivity, requirement of phosphatidic acid for activation, and by their activity toward native membranes (Boronenkov and Anderson, 1995).

PtdIns-3 kinase comprises a growing family of enzymes activated by a number of receptors including those of growth factors with intrinsic or associated tyrosine kinase activity or receptors coupled to trimeric G proteins (Kapeller and Cantley, 1994; Hunter, 1995). In mammalian cells, at least three biochemically distinct PtdIns-3 kinases have been recently described: the p110/p85 heterodimer, which associates through the SH2 domains in the p85 subunit with proteins that are tyrosine phosphorylated by protein tyrosine kinases (PTKs), a PtdIns-3 kinase activated by G-protein βγ subunits and a mammalian PtdIns-specific PtdIns-3 kinase. The term PI3-kinase will be used for the PTK/SH2 coupled form. This heterodimer comprises an 85 kDa regulatory and a 110 kDa cata-

lytic subunit and is the first reported incidence of a dual-specificity kinase that possesses both protein serine and lipid kinase activity (Hunter, 1995). PTK activity has a very limited range of substrates and appears to be largely confined to the autoregulation of the enzyme. In addition to biochemically distinct PI3-kinases, multiple forms exist within the p85 and p110 subunits. The p85 subunit contains a p110 binding site as well as several other important features such as two SH2 domains that mediate interaction with other signaling molecules, one SH3 domain and two proline-rich sequences which may also mediate protein-protein interactions. In addition, there is a region homologous to the C-terminal part of the Break-cluster region (BcR) gene product, the function of which is presently unknown.

Activation of PI3-kinase involves binding of p85 SH2 domains with specific phosphotyrosine-containing sequences in activated receptors (Fig. 3.3). This interaction promotes the relocalization of PI3-kinase to the plasma membrane and activation of the catalytic function of the associated p110 subunit. Specific tyrosine phosphorylated motifs are found in the cytoplasmic domains of receptor tyrosine kinases (RTKs), intracellular substrates such

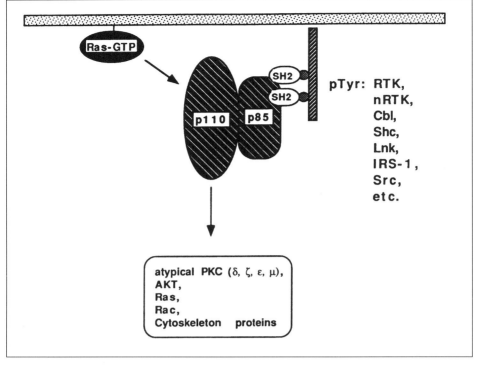

Fig. 3.3. PI3-kinase activation: possible pathways of PI3-kinase activation and major downstream target molecules.

as IRS-1 (insulin receptor substrate-1), Cbl and polyoma middle T antigen, and in cytosolic receptor-linked PTKs, such as c-Src. Binding of Src family kinases (including v-Src, Fyn and Lyn) via SH3 domains to the proline-rich regions on p85 has also been proposed to activate PI3-kinase in response to several stimuli (Kapeller and Cantley, 1994). Another mechanism by which PI3-kinase can be activated involves Ras which, in its activated GTP-bound state, can interact and associate directly with the p110 subunit of PI3-kinase leading to enhanced activation and increased production of 3-phosphorylated lipids. With respect to the Rho family GTPases, the involvement of Rho in the stimulation of PI3-kinase has been demonstrated by the use of the specific Rho inhibitor, C3 exoenzyme. Furthermore, association of Rac and Cdc42 with PI3-kinase has been recently reported, suggesting that these two small GTP-binding proteins may also regulate PI3-kinase activity (Tolias et al 1995).

PI3-kinase has been implicated in a variety of cellular processes including mitogenesis, cytoskeleton rearrangement and membrane traffic. PI3-kinase function may be related to its serine kinase activity, to the generation of D-3 phosphoinositide lipid products that act as second messengers as well as to its ability to act as an adaptor molecule. Thus, several biochemical events occur immediately downstream of PI3-kinase which may include a number of putative target molecules. The major effectors of PI3-kinase activity presently characterized include atypical PKC, such as ζ, δ, ϵ and η isozymes, small GTP-binding proteins, such as Ras and Rac, and Akt, a serine-threonine kinase involved in the activation of p70 S6 kinase (Ward et al 1996).

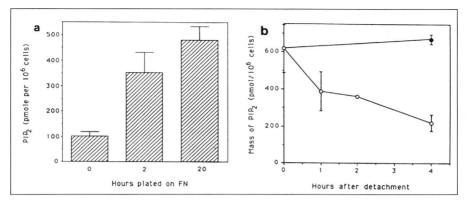

Fig. 3.4. (a) Attachment of C3H10T1/2 fibroblasts to fibronectin induces an increase in the mass of PtdIns(4,5)P$_2$. Cells were held in suspension for 18 h, then plated onto dishes coated with polyHEMA (poly-hydroxyethylmethacrylate) for 2 h (0 h on FN), or onto fibronectin for 2 h. Some cells were plated onto fibronectin at the start (20 h on FN). PtdIns(4,5)P$_2$ mass per cell was determined by measuring Ins(1,4,5,)P$_3$ concentration after alkaline hydrolysis of cellular lipids. The experiment was done in triplicate and values are means $\pm$ standard deviations. (b) Detachment of cells induces a decrease in the mass of PtdIns(4,5)P$_2$. Cells were trypsinized and held in suspension (O) or replated on fibronectin (●). At various times, levels of PtdIns(4,5)P$_2$ were determined as in (a). The point at T = 0 was taken immediately after trypsinization. The experiment was done in duplicate, and values are means $\pm$ standard deviation. Reproduced from McNamee HP et al J Cell Biol 1993; 121:673-678. By copyright permission of the Rockefeller University Press.

REGULATION OF PHOSPHOINOSITIDE METABOLISM BY INTEGRINS

Several lines of evidence indicate a relevant role for extracellular matrix components in the regulation of inositol lipid breakdown. A rapid activation of the PI turnover and increased levels of PtdIns(4)P and PtdIns(4,5)P$_2$ were first observed in BHK-21 cells spreading on fibronectin-coated culture plates by Breuer and Wagener (1989). Similarly, McNamee et al (1993) have reported that adhesion of C3H10T1/2 mouse fibroblasts on fibronectin or antibody-mediated integrin clustering induces a prompt increase in the phosphorylation of PtdIns(4)P to PtdIns(4,5)P$_2$ resulting in three- to five-fold increase in the absolute levels of cellular PtdIns(4,5)P$_2$ (Fig. 3.4). Change in PtdIns(4,5)P$_2$ is accompanied by modulation of the cellular responsiveness to PDGF, in that suspended cells with low PtdIns(4,5)P$_2$ show a decreased ability to generate Ins(1,4,5)P$_3$, activate PKC and

mobilize calcium in response to PDGF, as compared with adherent cells with high PtdIns(4,5)P$_2$ levels. Integrin-mediated increase in PtdIns(4,5)P$_2$ synthesis in mouse fibroblasts involves the activity of PtdIns-5 kinase, the enzyme that phosphorylates PtdIns(4)P to PtdIns(4,5)P$_2$ (Chong et al 1994). Enzyme activity and cellular levels of PtdIns(4,5)P$_2$ are regulated by the small GTP-binding protein Rho. Indeed, inactivation of small GTP-binding proteins by treatment with lovastatin induces a dramatic decrease in cellular levels of PtdIns(4,5)P$_2$, similar to loss of adhesion. Moreover, in cell lysates, activation or inactivation of Rho specifically triggers corresponding changes in the activity of PtdIns-5 kinase. Rho is also responsible for the changes in PDGF response observed in nonadherent cells. Thus, inactivation of Rho in botulinum C3 exoenzyme-treated cells leads to a decrease in PDGF-induced calcium mobilization and conversely microinjection of an activated variant of Rho re-

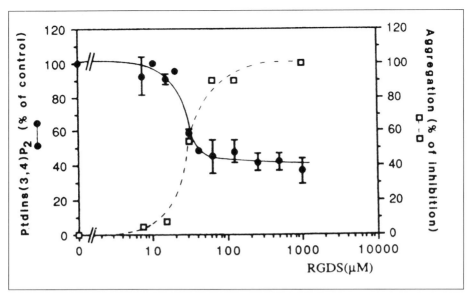

Fig. 3.5. Effects of tetrapeptide Arg-Gly-Asp-Ser (RGDS) on aggregation and PtdIns(3,4)P$_2$ synthesis in thrombin-stimulated platelets. Data (l) represent percentages of (^{32}P)PtdIns(3,4)P$_2$ radioactivity compared to controls without RGDS and are means ± S.E. of two to nine determinations. Under the conditions used, the radioactivity of (^{32}P)PtdIns(3,4)P$_2$ increased from nondetectable values in nonstimulated platelets to 1535 ± 233 cpm/10^8 cells in platelets stimulated with thrombin in the absence of RGDS. Platelet aggregation data (o) correspond to percentages of inhibition of aggregation compared to controls without RGDS. With permission from Sultan C et al J Biol Chem 1991; 266:23554-23557.

verses the decrease in calcium mobilization in nonadherent cells. These results strongly suggest that in nonadherent cells the decrease in PtdIns(4,5)P$_2$ synthesis is responsible for their diminished ability to respond to agonists that induce PtdIns(4,5)P$_2$ hydrolysis.

The finding that Rho, known primarily for its ability to promote cytoskeletal organization, is involved in the regulation of PtdIns(4,5)P$_2$ synthesis, and indirectly PtdIns(4,5)P$_2$ hydrolysis, raises the possibility that PtdIns(4,5)P$_2$ mediates some of the effects of Rho on the cytoskeleton. It can be further hypothesized that targeting of Rho or PtdIns-5 kinase to specific cellular locations such as focal adhesions may provide a means of regulating the sites of actin polymerization (see chapter 6).

With respect to the 3-phosphoinositide pathway, antibody-mediated crosslinking of

β2 integrins on neutrophils rapidly induces the formation of PtdIns(3,4,5)P$_3$ which correlates with the induction of actin polymerization (Lofgren et al 1993). Upon exposure to thrombin a marked accumulation of PtdIns(3,4) P$_2$, which is dependent on αIIbβ3 integrin binding to fibrinogen and partly requires tyrosine phosphorylation, has been reported in platelets. Using an Arg-Gly-Asp-Ser tetrapeptide, which blocks the interaction of fibrinogen with αIIbβ3, an inhibition of the late-phase of PtdIns(3,4)P$_2$ accumulation was found that was associated with increased Ca^{2+} (Fig. 3.5) (Sorisky et al 1992; Sultan et al 1991). In addition, pathological platelets from patients with Glazmann's thrombasthenia, which lack αIIbβ3 integrin and fail to aggregate in response to thrombin, displayed a barely detectable increase in PtdIns(3,4)P$_2$ synthesis (Sultan et al 1991). PtdIns(3,4)P$_2$ synthesis

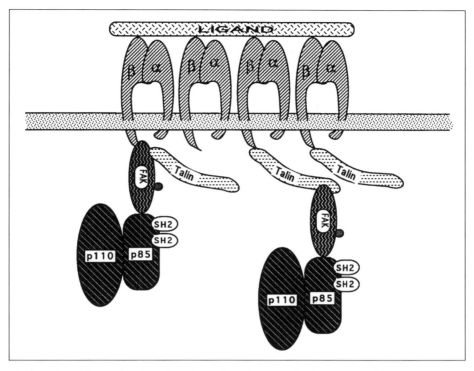

Fig. 3.6. Possible mechanisms leading to PI3-Kinase activation triggered by integrins.

parallels the translocation of both the p85 subunit of PI3-kinase and PI3-kinase activity to the cytoskeleton, in a TritonX-100 insoluble subcellular fraction which also contains actin, vinculin, talin, p60Src, PKC and αIIbβ3 (Zhang et al 1992). More recently, it has been demonstrated that integrin-dependent translocation of PI3-kinase to the cytoskeleton of thrombin-activated platelets involves specific interaction of p85 with actin filaments and tyrosine-phosphorylated p125Fak. This interaction is mediated by a proline-rich sequence (Lys-Pro-Pro-Arg-Pro-Gly) of human p125Fak which directly binds to the SH3 domain of p85 (Guinebault et al 1995). Moreover, a peptide corresponding to the proline rich region of p125Fak increases the specific activity of PI3-kinase suggesting that interaction of p125Fak with the p85 subunit of PI3-kinase may represent a new mechanism of PI3-kinase activation (Fig. 3.6). An interaction of PI3-ki-

nase with p125Fak has been also observed in fibroblasts upon cell adhesion to fibronectin (Fig. 3.7). This study indicates that p125Fak autophosphorylation is critical for its binding to PI3-kinase and that the p85 subunit of PI3-kinase is a substrate of p125Fak in vitro (Chen and Guan, 1994a).

Overall, these data suggest that p125Fak may be responsible, at least in part, for the indirect binding of PI3-kinase to actin filaments and plays a critical role in activating and directing this lipid kinase at very specific areas such as focal contact points.

p125Fak/PI3-kinase interaction has also been described in NIH3T3 mouse fibroblasts in response to PDGF, but differently regulated as compared with that induced by cell adhesion (Chen and Guan, 1994b). Indeed, stimulation of p125Fak/PI3-kinase association induced by cell adhesion to fibronectin occurs concomitantly to p125Fak phosphorylation and requires the

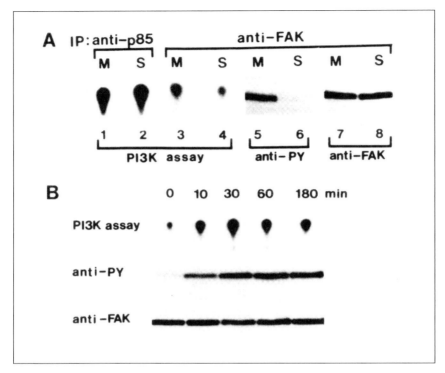

Fig. 3.7. Effect of tyrosine phosphorylation of Fak on the association of PI3-kinase activity. (A) Equal amounts of lysates from attached monolayer (M) or suspended (S) NIH 3T3 cells were immunoprecipitated (IP) with anti-p85 (lanes 1 and 2) or anti-Fak (lanes 3-8). The anti-p85 immunoprecipitates were subjected to PI3-kinase (PI3-K) assays (lanes 1 and 2). Aliquots of the anti-Fak immunoprecipitates were assayed for PI3-kinase assay (lanes 3 and 4), or separated by SDS-PAGE and Western blotted with antiphosphotyrosine (pY) (lanes 5 and 6) or with anti-Fak (lanes 7 and 8) antibodies, respectively. (B) Suspended NIH 3T3 cells (0 min) were plated onto culture dishes that had been coated with fibronectin (10 μg/ml) and blocked by bovine serum albumin. At 10, 30, 60, and 180 min after plating, lysates were prepared and immunoprecipitated by anti-Fak. They were divided into three parts and used for assays as described in A. With permission from Chen HC and Guan JL Proc Natl Acad Sci USA 1994; 91:10148-10152. Copyright 1994 National Academy of Sciences, USA.

integrity of the actin cytoskeleton. In contrast, PDGF-induced p125Fak/PI3-kinase association does not correlate with the level of p125Fak phosphorylation in vivo and is not affected by disruption of the actin cytoskeleton. These results provide another possible mechanism of crosstalk between the signaling pathways initiated by growth factors and integrins and raise the possibility that p125Fak may participate in the growth factor effect by modulating cell morphology and migration.

PHOSPHOLIPASES

Receptor-mediated hydrolysis of cellular phospholipids is a ubiquitous biochemical event of central importance in cell signal transduction. This hydrolysis is catalyzed by receptor-linked phospholipases of distinct specificities (PLC, PLD, PLA_2) and gives rise to multiple, sometimes interconvertable, lipid-derived second messengers (Fig. 3.8 and Table 3.1) (Liscovitch, 1992). Among the phospholipids that are substrates for signal-activated phospholi-

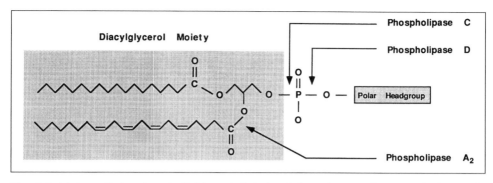

Fig. 3.8. Major cellular phospholipids and their breakdown by signal-activated phospholipases. Glycerophospholipids consist of a hydrophobic DAG moiety, linked via a phosphodiester bond to a polar headgroup which differ among phospholipid classes. The covalent bonds cleaved by the specific phospholipases are indicated.

pases are minor membrane constituents such as PtdIns(4,5)P_2 as well as more abundant phospholipid species, such as phosphatidylcholine (PtdCho), phosphatidylethanolamine (PtdEtn) and sphingomyelin.

A class of PLC (Ptd-PLC) acts on PtdIns(4,5)P_2 to generate DAG and Ins(1,4,5)P_3. Inositol(1,4,5)trisphosphate mobilizes Ca^{2+} from intracellular stores and DAG activates PKC. Like many other enzymes involved in signal transduction, it exists in multiple isoforms. All PLC enzymes identified thus far can be divided into three types, γ, β and δ (Lee and Rhee, 1995).

PLCγ1 is activated via several growth factor RTKs. Growth factor-induced activation of PLCγ1 requires its tyrosine phosphorylation as well as association of the enzyme with the RTK. Nonreceptor PTKs also phosphorylate the same tyrosine residues as the RTKs thus activating this enzyme in response to ligation of certain cell surface receptors, including the T-cell antigen receptor, the membrane associated IgM, the high-affinity receptor for IgE, the IgG receptors and several cytokine receptors.

PLCβ isozymes are activated by the GTP-bound α subunit of the G_q class of the heterotrimeric GTP-binding proteins. The receptors that utilize this pathway include those for thromboxane A$_2$, bradykinin, bombesin, histamine, vasopressin, acetyl-

choline, etc. PLCβ isozymes are also activated by the βγ subunits of G proteins; the PLCβ region that interacts with Gβγ dimers differs from that responsible for interaction with G_qα subunit suggesting that PLCβ isozymes can be activated by G_qα and Gβγ subunits in an additive manner.

The activation mechanisms of PLCδ isoenzyme are not known at the present time.

Another distinct class of PLC appears to act exclusively on PtdCho, the most abundant phospholipid species of the mammalian membrane, to produce DAG and phosphocholine (Exton 1994).

PLD can also degrade PtdCho in numerous receptor-mediated systems, generating phosphatidic acid (PtdOH), a molecule with potential second messenger functions, and choline (Table 3.1). PtdOH is further metabolized by PtdOH phosphohydrolase to form DAG. PLD activities exist in both membrane-bound and cytosolic forms, reflecting the existence of different isoenzymes. Membrane-associated activities exhibit strict specificities for PtdCho, whereas the cytosolic forms hydrolyze PtdEtn and PtdIns. Activation of PLD occurs through multiple mechanisms involving PKC, tyrosine kinases, Ca^{2+} and small GTP-binding proteins (Billah 1993).

Table 3.1. Products of enzymatic hydrolysis of common glycerophosphatides

Enzyme	Lipid	Products
Phospholipase A_2	Phosphatidic acid Phosphatidyl choline Phosphatidyl ethanolamine Phosphatidyl serine Phosphatidyl glycerol	sn-1-Acyl-3-glycerophosphate sn-1-Acyl-3-glycerophosphoryl-choline sn-1-Acyl-3-glycerophosphoryl-ethanolamine + fatty acid sn-1-Acyl-3-glycerophosphoryl-serine sn-1-Acyl-3-glycerophosphoryl-glycerol
Phospholipase C	Phosphatidyl choline Phosphatidyl ethanolamine Phosphatidyl glycerol 1-lysophosphatidyl choline PtdIns(4,5)P_2	sn-1,2-Diglyceride + phosphorylcholine sn-1,2-Diglyceride + phosphorylethanolamine sn-1,2-Diglyceride + sn-1-glycerophosphate sn-1-Monoacylglycerol + phosphorylcholine sn-1,2-Diglyceride + Ins(1,4,5)P_3
Phospholipase D	Phosphatidyl choline Phosphatidyl ethanolamine Phosphatidyl serine Phosphatidyl glycerol 1-lysophosphatidyl choline	sn-1,2-Diacyl-3-glycerophosphate + choline sn-1,2-Diacyl-3-glycerophosphate + ethanolamine sn-1,2-Diacyl-3-glycerophosphate + serine sn-1,3-Diacylglycerophosphate + glycerol sn-1-Acyl-3-glycerophosphate + choline

PLA$_2$ catalyzes the hydrolysis of the sn-2 fatty acyl-bond of phospholipids to liberate free fatty acid and lysophospholipids (Table 3.1) (Dennis, 1994). The major sources of arachidonic acid (AA) and its metabolites released by agonist activation of PLA$_2$ are PtdCho, PtdEtn and PtdIns. PLA$_2$s are a diverse class of enzymes in regard to function, localization and regulation. The molecular characteristics of three mammalian PLA$_2$ isoforms, referred to as Groups I, II and IV, have been recently elucidated. Groups I and II are the counterparts of types I and II snake venom secretory PLA$_2$ (sPLA$_2$); group IV is a receptor coupled cytosolic PLA$_2$ (cPLA$_2$). Secretory and cytosolic PLA$_2$ differ in a number of aspects including preferential activity for AA-containing phospholipids and different sensitivity to reducing agents and specific pharmacological inhibitors. In addition, they have different requirements in terms of calcium concentration for optimal activity. The major polyunsaturated free fatty acid generated upon PLA$_2$ activation is AA, the rate-limiting precursor of leukotriene and prostaglandin synthesis.

Several observations indicate that PL-generated lipid mediators may affect cytoskeleton organization (Gips et al 1994). DAG, independent of PKC stimulation, enhances the activity of a nucleating protein for actin filaments. DAG can also, through the activation of PKC, modify specific interactions of actin with actin-binding proteins such as MARCKS. PtdOH has been shown to be a key molecule for the stabilization of actin stress fiber. Arachidonate and its metabolism products have been found to regulate the actin cytoskeleton in response to growth factors.

PLCγ ACTIVATION THROUGH INTEGRINS

Few data are available on integrin receptor stimulation of PLCγ activation and hydrolysis of phosphoinositides.

The first results indicating that integrin ligation results in Ins(1,4,5)P$_3$ generation were reported in lymphoid cells. Crosslinking of the α, but not β, subunit of LFA-1 re-sulted in phosphoinositide hydrolysis and a rise of intracellular calcium level in both T and NK cells (Pardi et al 1989). Tyrosine phosphorylation of PLCγ1, which is completely inhibited by the PTK inhibitor herbimycin A, has been described following antibody-mediated β2-integrin ligation on CD4$^+$ T cells (Fig. 3.9) (Kanner et al 1993). Moreover, β2 integrin crosslinking induces tyrosine phosphorylation of an additional pp35/36 protein, which associates with the SH2 domains of PLCγ and Grb2 after T cell receptor (TCR) stimulation and has been suggested to play a critical role in coupling the TCR with the phosphatidylinositol second-messenger pathway (Sieh et al 1994; Motto et al 1996). It has also been reported that tyrosine phosphorylation of PLCγ is prolonged when TCR and β2 integrins are simultaneously engaged with antibody and the counter receptor ICAM-1, respectively, in comparison to ligation of TCR/CD3 alone (Kanner et al 1993).

With respect to the ability of β1 integrin engagement to stimulate PtdIns-specific PLCγ in T cells, Sato et al (1995) have reported that solid phase crosslinking of α4β1 with anti-α4 monoclonal antibodies on resting T lymphocytes induces tyrosine phosphorylation of several cellular substrates including PLCγ1. Unlike ICAM-1, however, the cellular ligand of α4β1, VCAM-1, does not prolong the TCR-triggered tyrosine phosphorylation of PLCγ1 (Kanner et al 1993).

Extracellular matrix components are also able to stimulate PLCγ activation. Thus, platelet stimulation by collagen results in generation of Ins(1,4,5)P$_3$ (Daniel et al 1994) which is accompanied by the tyrosine phosphorylation of PLCγ2, but not PLCγ1 (Daniel et al 1994; Blake et al 1994). However, the involvement of α2β1 integrin in collagen-induced activation of PLCγ has not been specifically addressed and several platelet surface proteins with collagen binding properties are potentially involved.

Collagen has been described to stimulate PLC activation also in other cell types. Adhesion of rat glomerular epithelial cells to collagen type I or IV results in calcium-

Fig. 3.9. PLC-γ1 tyrosine phosphorylation in response to CD18 stimulation. T-cell blasts were incubated in the absence (-) or presence (+) of herbimycin A (1 μg/ml) for 16 hour and then stimulated with either anti-CD18 or anti-TCR. Immunoprecipitates of PLC-γ1 were then prepared and immunoblotted with anti-pTyr (upper) or anti-PLC-γ1 (lower). With permission from Kanner SB et al Proc Natl Acad Sci USA 1993; 90:7099-7103. Copyright 1993 National Academy of Sciences, USA.

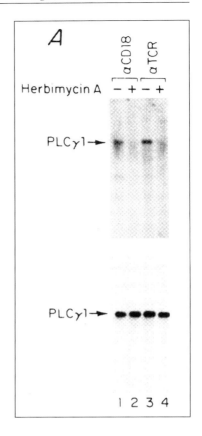

dependent increase in DAG and inositol phosphates and decrease in inositol phospholipids, which are partially inhibited in the presence of a Fab fragment of an anti-β1 integrin antiserum (Cybulski et al 1990 and 1993). In contrast to Fab fragments, intact rabbit anti-β1 antibodies enhanced the DAG production mimicking the crosslinking of integrins by substratum-bound collagen. Likewise, Somogyi et al (1994) have reported tyrosine kinase-dependent stimulation of inositol lipid breakdown by collagen type IV in pancreatic acinar cells, by direct mass measurement of inositol lipids.

PLA₂ ACTIVATION THROUGH INTEGRINS

The first indication that adhesion to ECM components results in the activation of PLA₂ was provided by Chun and Jacobson (1992), who demonstrated that the interaction of HeLa cells with immobilized gelatin or RGD peptide activates PLA₂ to release free AA (Fig. 3.10). Their data also show that the inhibition of PLA₂ or lipoxygenase blocks HeLa cell spreading suggesting that lipoxygenase metabolite(s) formed from AA upon collagen receptor clustering may function as second messenger(s) that initiates cell spreading. The lipoxygenase inhibitors also block the spreading of other cells types including endothelial cells and fibroblasts on gelatin or fibronectin. This finding further supports the hypothesis that this second messenger system regulates the adhesion of various cells to different substrata.

Generation of AA has been also reported in rat glomerular epithelial cells upon adhesion to collagen (Cybulski et al 1993). Col-

lagen-induced increase in free AA is partially inhibited by lowering extracellular calcium concentrations or by a Fab fragment of an anti-β1 antibody.

Direct evidence that β1 integrins are capable of initiating a PLA₂-mediated signaling pathway was provided in a subsequent study by Auer and Jacobson (1995) showing that antibody-mediated ligation of β1 integrins on Hela cells results in PLA₂-dependent AA release.

As mentioned above, PLA₂ comprise a family of different enzymes. Thus far, no data are available on the molecular characteristics of PLA₂ enzymes activated upon integrin clustering. The ability of bromo phenacylbromide, a specific inhibitor of secretory PLA₂, to block integrin-dependent AA generation (Auer and Jacobson, 1995) strongly suggests that increased enzymatic

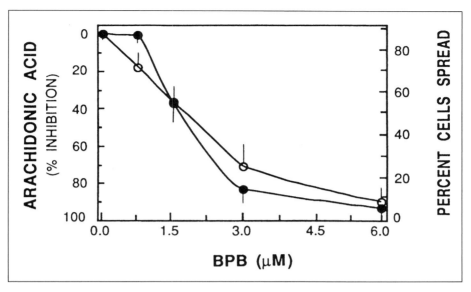

Fig. 3.10. Dose-dependent effects of the PLA2 inhibitor BPB (bromophenacyl bromide) on cell spreading (●) and arachidonic acid (AA) release (°). HeLa cells were treated with indicated amounts of BPB for 5 min in suspension, plated on gelatin at 37°C for 30 min and percentage of cell spread was scored. To determine AA release, (^{3}H)AA-labeled cells were treated with BPB and plated on gelatin at 37°C. The amount of (^{3}H)AA was determined after 20-min incubation in which maximum amount of AA is released. The data represent the average of three experiments and the error bars indicate the standard deviation. Reproduced from Chun JS and Jacobson B-S Mol Biol Cell 1992; 3:481-492, with permission of the American Society for Cell Biology.

activity is mediated by this class of enzymes. On the other hand, the activation of cytosolic PLA_2 might also occur given that cytosolic PLA_2 is a major substrate of Erk kinases (Lin et al 1993), which are activated through integrin receptors (see chapter 5).

PLD ACTIVATION THROUGH INTEGRINS

Only a few reports are available on the ability of integrins to induce PLD activity. Stimulation of human neutrophils with the complement component C3bi results in an increased accumulation of diglyceride and in a rapid and significant breakdown of PtdCho. The involvement of PLD activity in these events has been demonstrated by the ability of C3bi to induce an increased formation of alkyl-PtdOH and alkyl-diglyceride in 3H-alkyl-lyso-PtdCho labeled cells and by a marked accumulation of alkyl-

phosphatidylethanol, when stimulation of these prelabeled cells was performed in the presence of ethanol (Fallman et al 1992).

Similarly, fibroblast contraction of stressed collagen matrices results in a rapid burst of PtdOH release. Also in this case, PLD activity has been implicated in production of PtdOH based on the observation of a transphosphatidylation reaction in the presence of ethanol that resulted in formation of phosphatidylethanol at the expense of PtdOH. Activation of PLD required extracellular calcium ions and was regulated by PKC (He and Grinnell, 1995).

PROTEIN KINASES C

PKC are related serine and/or threonine protein kinases that transduce a number of signals promoting lipid hydrolysis (Newton, 1995). The occupancy of a variety of membrane receptors, including G protein-

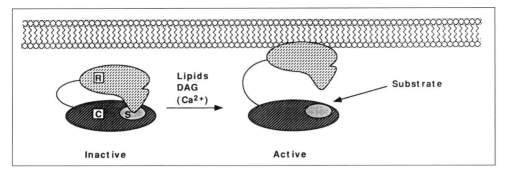

Fig. 3.11. Model for activation of PKC. A representative PKC isotype is depicted with its modular structure (C, catalytic domain; R, regulatory domain; S, substrate-binding pocket). The binding of the regulatory domain to the substrate-binding pocket maintains the inactive conformation. Lipids, DAG and Ca^{2+} can activate PKC leading to dissociation of the regulatory domain from the substrate-binding pocket which can now bind and phosphorylate specific substrates.

coupled receptors, RTKs or nonreceptor tyrosine kinases, can cause the generation of PKC activator, DAG, either rapidly by activation of specific PLCs or more slowly by activation of PLD to yield PtdOH and then DAG. In addition, fatty acid and lysoPtdCho generation by PLA_2 activation modulates PKC activity (Fig. 3.11). Phorbol esters may substitute for DAG in activating PKC and treatment of cells with these molecules results in prolonged activation of PKC. Eleven different PKC isozymes have been described to date which display both common features and substantial differences in primary structure, enzymatic activities, tissue and intracellular distribution, and cofactor requirement (Table 3.2) (Hug and Sarre, 1993; Dekker and Parker, 1994).

Based on the activation mechanisms, PKC have been classified in four groups: Ca^{2+}-dependent or conventional PKC-α, -β and -γ; Ca^{2+}-independent novel PKC-δ, -ε, -θ and -η; atypical PKC-ζ, -λ and -ι; and PKC-μ. In addition to regulation by DAG or phorbol esters, all PKC isozymes require phosphatidylserine (PtdSer), an acidic lipid located exclusively on the cytoplasmic face of membranes, and Ca^{2+} for optimal activity. Moreover, recent evidence indicates that $PtdIns(3,4)P_2$ and $PtdIns(3,4,5)P_3$, both produced in agonist-stimulated cells, are potent and selective activators of PKC-δ, -ε, -η and -ζ suggesting that PI3-Kinase is an upstream molecule regulating atypical PKC activity.

MARCKS (myristoylated alanin-rich C-kinase substrates) are included among the physiologically relevant substrates of PKC. These are widely distributed acidic proteins binding both calmodulin and actin, which have been implicated in a variety of biological processes including secretion, membrane trafficking, cell motility, phagocytosis, regulation of the cell cycle and transformation. All of these processes necessitate regulated rearrangement of the actin cytoskeleton. Recent data suggest that MARCKS play a role in translating extracellular signals into alterations in actin plasticity and changes in actin-membrane interaction (Aderem, 1992).

PKC ACTIVATION THROUGH INTEGRINS

The involvement of PKC in cell adhesion to extracellular matrix components has been suggested from the observation that PKC-activating phorbol esters enhance spreading and focal adhesion formation in various cell types in response to the extracellular matrix components and that PKC inhibitors prevent them (Woods and Couchman, 1992; Chun and Jacobson, 1992). Moreover, evidence has been provided for the association of PKC-α with focal contacts in rat embryo fibroblasts (Jaken et al 1989) and for the re-

Table 3.2. PKC isoforms in mammalian tissues

	Subspecies	Ca^{2+} and Lipid activators	Tissue expression
cPKC	α	Ca^{2+}, DAG, PtdSer, FFAs, LysoPtdCho	universal
	βI	Ca^{2+}, DAG, PtdSer, FFAs, LysoPtdCho	some tissues
	βII	Ca^{2+}, DAG, PtdSer, FFAs, LysoPtdCho	many tissues
	γ	Ca^{2+}, DAG, PtdSer, FFAs, LysoPtdCho	brain only
nPKC	δ	DAG, PtdSer, PtdIns(3,4,5)P$_3$	universal
	ε	DAG, PtdSer, FFA, PtdIns(3,4,5)P$_3$, PtdIns(3,4)P$_2$	brain and others
	η	DAG, PtdSer, PtdIns(3,4,5)P$_3$, PtdIns(3,4)P$_2$, Cholesterol sulfate	skin, lung, heart
	θ	?	muscle, T cell, etc.
	μ	?	NRK cells
aPKC	ζ	PtdSer, FFA, PtdIns(3,4,5)P$_3$?, PtdIns(3,4)P$_2$	universal
	λ	?	many tissues

The activators for each isoform are determined with calf thymus H1 histone and bovine myelin basic protein as model phosphate acceptors. FFA, cis-unsaturated fatty acid.

cruitment of PKC-δ to focal adhesions in serum-stimulated 3T3 cells (Barry and Critchley, 1994). Direct evidence of PKC activation has been reported in Chinese hamster ovary (CHO) cells (Fig. 3.12) and human monocytes in response to fibronectin (Vuori and Ruoslahti, 1993; Chang et al 1993) and in HeLa cells plated on a collagen substratum (Chun and Jacobson 1993). In HeLa cells, adhesion to collagen induces rapid release of DAG followed by an increase in membrane-bound PKC that occurs during the attachment phase of cell adhesion and prior to cell spreading. In CHO cells plated on fibronectin, PKC activity transiently increases in the membrane fraction before cell spreading. PKC activity is required for fibronectin-induced cell spreading as well as for tyrosine phosphorylation of p125Fak; PKC-α, however, has not been found to act directly on p125Fak (Vuori and Ruoslahti, 1993). The involvement of PKC in p125Fak activation was also demonstrated in platelets adhering to fibrinogen via αIIbβ3 integrin (Haimovich et al 1996).

Stimulation of human monocytes with the 120 kD fragment of fibronectin triggers a rapid and brief activation of PKC which is required for fibronectin-induced TNF secretion (Chang et al 1993) and in neutrophils following β2 integrin stimulation (Fallman et al 1992). In regard to the PKC isozymes activated in response to cell adhesion to extracellular matrix components, a recent report by Chun et al (1996) has demonstrated redistribution of PKC-ε, but not of the other isozymes (α, γ, ζ, λ and ι) expressed in HeLa cells, from the cytosol to the membrane during cell attachment to a gelatin substratum (Fig. 3.13).

PKC may also regulate integrin functions through phosphorylation of the PKC-specific substrate MARCKS, a phosphoprotein that localizes to focal contact-like sites. In this regard, a role for MacMARCKS, a member of the MARCKS family, has been recently reported in integrin-dependent signal transduction pathway in macrophages (Li et al 1996). In this system, a dominant negative mutant of MacMARCKS was found

Fig. 3.12. Measurement of protein kinase C activity. (A), CHO cells were plated on either fibronectin- or polylysine-coated dishes and protein kinase C activity in membrane fractions was measured at the indicated times. The kinase activity is expressed as picomoles of phosphate incorporated into a synthetic substrate per min. (B), cells were plated on the same substrates as in A for 12 min, and protein kinase C activity in cytosolic and membrane fractions was determined. Results are expressed as percent of total activity. With permission from Vuori K and Ruoslahti E J Biol Chem 1993; 268:21459-21462.

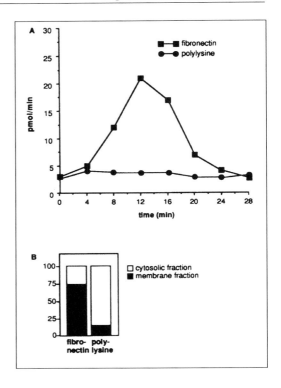

Fig. 3.13. Translocation of PKCE during HeLa cell adhesion to a gelatin substratum. HeLa cells were either kept in suspension (0 min) or plated on a gelatin substratum for the indicated periods. Cells were fractionated, and the lysates (25 μg) were used for immunoblotting with PKC isozyme-specific antibodies. C, cytosolic fraction; M, membrane fraction; and P, particulate fraction. With permission from Chun JS et al J Biol Chem 1996; 271:13008-13012.

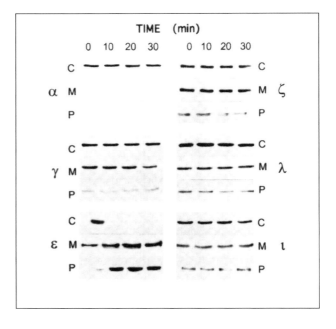

to block PMA and immunocomplex-induced macrophage spreading as well as integrin-dependent tyrosine phosphorylation of paxillin. Furthermore MacMARCKS colocalizes with paxillin in the membrane ruffles at the leading edge of the spreading cells, providing a potential site for MacMARCKS to participate in the regulation of paxillin phosphorylation.

Overall, these findings suggest that activation of multiple forms of PKC plays a key role in integrin-dependent phosphorylation of specific proteins (i.e. p125Fak, paxillin), cell spreading and other cellular functions.

REFERENCES

Aderem A (1992) Signal transduction and the actin cytoskeleton: the roles of MARCKS and profilin. Trends Biochem Sci 17:438-443.

Auer KL, Jacobson BS (1995) β1 integrins signal lipid second messengers required during cell adhesion. Mol Biol Cell 6:1305-1313.

Barry ST, Critchley DR (1994) The RhoA-dependent assembly of focal adhesions in Swiss 3T3 cells is associated with increased tyrosine phosphorylation and the recruitment of both pp125FAK and protein kinase C-δ to focal adhesions. J Cell Sci 107:2033-2045.

Billah MM (1993) Phospholipase D and cell signaling. Curr Opin Immunol 5:114-123

Blake RB, Schieven GL, Watson SP (1994) Collagen stimulates tyrosine phosphorylation of phospholipase C-γ2 but not phospholipase C-γ1 in human platelets. FEBS Lett 353:212-216.

Boronenkov IV, Anderson RA (1995) The sequence of phosphatidylinositol-4-phosphate 5-kinase defines a novel family of lipid kinases. J Biol Chem 270:2881-2884

Breuer D, Wagener C (1989) Activation of the phosphatidylinositol cycle in spreading cells. Exp Cell Res 182:659-663.

Carpenter CL, Cantley LC (1990) Phosphoinositide kinases. Biochem 29:11147-11156.

Chang ZL, Beezhold DH, Personius CD, Shen ZL (1993) Fibronectin cell-binding domain triggered transmembrane signal transduction in human monocytes. J Leukoc Biol 53:79-85.

Chen HC, Guan JL (1994a) Association of focal adhesion kinase with its potential substrate phosphatidylinositol 3-kinase. Proc Natl Acad Sci USA 91:10148-10152.

Chen HC, Guan JL (1994b) Stimulation of phosphatidylinositol 3'-kinase association with focal adhesion kinase by platelet-derived growth factor. J Biol Chem 269:31229-31233.

Chong LD, Traynor-Kaplan A, Bokoch GM, Schwartz MA (1994) The small GTP-binding protein Rho regulates a phosphatidylinositol 4-phosphate 5-kinase in mammalian cells. Cell 79:507-513.

Chun JS, Ha MJ, Jacobson BS (1996) Differential translocation of protein kinase C ε during HeLa cell adhesion to a gelatin substratum. J Biol Chem 271:13008-13012.

Chun JS, Jacobson BS (1992) Spreading of HeLa cells on collagen substratum requires a second messenger formed by the lipoxygenase metabolism of arachidonic acid released by collagen receptor clustering. Mol Biol Cell 3:481-492.

Chun JS, Jacobson BS (1993) Requirement for diacylglycerol and protein kinase C in HeLa cell-substratum adhesion and their feedback amplification of arachidonic acid production for optimum cell spreading. Mol Biol Cell 4:271-281.

Cybulsky AV, Bonventre JV, Quigg RJ, Wolfe LS, Salant DJ (1990) Extracellular matrix regulates proliferation and phospholipid turnover in glomerular epithelial cells. Am J Physiol 259:F326-F337.

Cybulsky AV, Carbonetto S, Cyr M-D, McTavish AJ, Huang Q (1993) Extracellular matrix-stimulated phospholipase activation is mediated by β1-integrin. Am J Physiol 264:C323-C332.

Daniel JL, Dangelmaier C, Smith JB (1994) Evidence for a role for tyrosine phosphorylation of phospholipase Cγ2 in collagen-induced platelet cytosolic calcium mobilization. Biochem J 302:617-622.

Dekker LV, Parker PJ (1994) Protein kinase C- a question of specificity. Trends Biochem Sci (TIBS) 19:73-77.

Dennis EA (1994) Diversity of group types, regulation, and function of phospholipase A₂. J Biol Chem 269:13057-13060.

Divecha N, Irvine RF (1995) Phospholipid signaling. Cell 80:269-278.

Exton JH (1994) Phosphatidylcholine breakdown and signal transduction. Biochim Biophys Acta 1212:26-42.

Fallman M, Gulbert M, Hellberg C, Andersson T (1992) Complement receptor-mediated phagocytosis is associated with accumulation of phosphatidylcholine-derived diglyceride in human neutrophils. J Biol Chem 267:2656-2663.

Gips SJ, Kandzari DE, Goldschmidt-Clermont PJ (1994) Growth factor receptors, phospholipases, phospholipid kinases and actin reorganization. Sem Cell Biol 5:201-208.

Guinebault C, Payrastre B, Racaud-Sultan C, Mazarguil H, Breton M, Mauco G, Plantavid M, Chap H (1995) Integrin-dependent translocation of phosphoinositide 3-kinase to the cytoskeleton of thrombin-activated platelets involves specific interaction of p85α with actin filaments and focal adhesion kinase. J Cell Biol 129:831-842.

Haimovich B, Kaneshiki N, Ji P (1996) Protein kinase C regulates tyrosine phosphorylation of pp125FAK in platelets adherent to fibrinogen. Blood 87:152-161.

Harlan JE, Hajduk PJ, Yonn HS, Fesik SW (1994) Pleckstrin homology domains bind to phosphatidylinositol-4,5-bisphosphate. Nature 371:168-170.

He Y, Grinnell F (1995) Role of phospholipase D in the cAMP signal. Transduction pathway activated during fibroblast contraction of collagen matrices. J Cell Biol 130:1197-1205.

Hug H, Sarre TF (1993) Protein kinase C isoenzymes: divergence in signal transduction? Biochem J 291:329-343.

Hunter T (1995) When is a lipid kinase not a lipid kinase? When it is a protein kinase. Cell 83:1-4.

Jaken S, Leach K, Klauck T (1989) Association of type 3 protein kinase C with focal contacts in rat embryo fibroblasts. J Cell Biol 109:697-704.

Kanner SB, Grosmaire LS, Ledbetter JA, Damle NK (1993) β2-integrin LFA-1 signaling through phospholipase C-γ1 activation. Proc Natl Acad Sci USA 90:7099-7103.

Kapeller R, Cantley LC (1994) Phosphatidylinositol 3-kinase. Bioassays 16:565-576.

Lee SB, Rhee SG (1995) Significance of PIP2 hydrolysis and regulation of phospholipase C isozymes. Curr Opin Cell Biol 7:183-189.

Li J, Zhu Z, Bao Z (1996) Role of MacMARCKS in integrin-dependent macrophage spreading and tyrosine phosphorylation of paxillin. J Biol Chem 271:12985-12990.

Lin LL, Wartmann M, Lin AY, Knopf JL, Seth A, Davis R (1993) Cytosolic phospholipase A₂ is phosphorylated and activated by MAP kinase. Cell 72:269-278.

Liscovitch M (1992) Crosstalk among multiple signal-activated phospholipases. Trends Biochem Sci 17:393-399.

Logfgren R, Ng-Sikorski J, Sjolander A, Andersson T (1993) β2 integrin engagement triggers actin polymerization and phosphatidylinositol trisphosphate formation in nonadherent human neutrophils. J Cell Biol 123:1597-1605.

McNamee HP, Ingberg DE, Schwartz MA (1993) Adhesion to fibronectin stimulates inositol lipid synthesis and enhances PDGF-induced inositol lipid breakdown. J Cell Biol 121:673-678.

Motto DG, Musci MA, Ross SE, Koretzky GA (1996) Tyrosine phosphorylation of Grb2-associated proteins correlates with phospholipase Cγ1 activation in T cells. Mol Cell Biol 16:2823-2829.

Newton AC (1995) Protein kinase C: structure, function, and regulation. J Biol Chem 270:28495-28498.

Pardi R, Bender JR, Dettori C, Giannazza E, Engleman EG (1989) Heterogeneous distribution and transmembrane signaling properties of lymphocyte function-associated antigen (LFA-1) in human lymphocyte subsets. J Immunol 143:3157-3166.

Sato T, Tachibana K, Nojima Y, D'Avirro N, Morimoto C (1995) Role of the VLA-4 molecule in T cell costimulation. Identification of the tyrosine phosphorylation pattern induced by the ligation of VLA-4. J Immunol 155:2938-2947.

Sieh M, Batzer A, Schlessinger J, Weiss A (1994) Grb2 and phospholipase C-γ1 associate with a 36- to 38-kilodalton phosphotyrosine protein after T-cell receptor stimulation. Mol Cell Biol 14:4435-4442.

Somogyi L, Lasic Z, Vukicevic S, Banfic H (1994) Collagen type IV stimulates an increase in intracellular Ca^{2+} in pancreatic acinar cells via activation of phospholipase C. Biochem J 299:603-61.

Sorisky A, King WG, Rittenhouse SE (1992) Accumulation of PtdIns(3,4)P_2 and PtdIns(3,4,5)P_3 in thrombin-stimulated platelets. Different sensitivities to Ca^{2+} or functional integrin. Biochem J 286:581-584.

Sultan C, Plantavid M, Bachelot C, Grondin P, Breton M, Mauco G, Levy-Toledano S, Caen JP, Chap H (1991) Involvement of platelet glycoprotein IIb-IIIa (αIIb-β3 integrin) in thrombin-induced synthesis of phosphatidylinositol 3',4'-bisphosphate. J Biol Chem 266:23554-23557.

Tolias KF, Cantley LC, Carpenter CL (1995) Rho family GTPases bind to phos-phoinositide kinases. J Biol Chem 270:17656-17659.

Vuori K, Ruoslahti E (1993) Activation of protein kinase C precedes α5β1 integrin-mediated cell spreading on fibronectin. J Biol Chem 268:21459-21462.

Ward SG, June CH, Olive D (1995) PI3-kinase: a pivotal pathway in T-cell activation? Immunol Today 17:187-197.

Woods A, Couchman JR (1992) Protein kinase C involvement in focal adhesion formation. J Cell Sci 101:277-290.

Zhang J, Fry MJ, Waterfield MD, Jaken S, Liao L, Fox JEB, Rittenhouse SE (1992) Activated phosphoinositide 3-kinase associates with membrane skeleton in thrombin-exposed platelets. J Biol Chem 267:4686-4692.

CHAPTER 4

Intracellular Calcium, pH and Membrane Potential

MECHANISMS OF INTRACELLULAR CA^{2+} CONTROL

Calcium ions, Ca^{2+}, are widely used second messengers in signaling systems (Clapham, 1995). Ca^{2+}-binding proteins undergo allosteric changes and modify their activity upon Ca^{2+} binding. The best known molecules with these properties are troponin C and calmodulin. The first regulates skeletal muscle contraction by modifying actin-myosin interactions. Calmodulin is an ubiquitous and abundant molecule that regulates the activity of different cytoplasmic proteins.

In resting conditions cytoplasmic levels of Ca^{2+} are in the range of 100 nM, while the extracellular calcium concentration is 1-2 x 10^4-fold higher. Ca^{2+} is also stored at high concentration in the endoplasmic reticulum inside the cells. The low Ca^{2+} level in the cytoplasm is maintained by calcium pumps that, by consuming ATP, transfer Ca^{2+} ions from the cytoplasm to either the extracellular space or into the endoplasmic reticulum (Schatzmann, 1989). Thus, Ca^{2+} levels in the cytoplasm are regulated mainly by the opening and closing of Ca^{2+} channels present both in the endoplasmic reticulum and in the plasma membrane.

Distinct mechanisms control channel opening in these two compartments. The release of Ca^{2+} from the endoplasmic reticulum stores occurs via a Ca^{2+} channel whose opening is triggered by the inositol 1,4,5-trisphosphate (Ins(1,4,5)P$_3$) (Berridge, 1993). This molecule is generated from the hydrolysis of phosphatidylinositol (4,5) bisphosphate (PtdIns(4,5)P$_2$) through the action of two known PtdIns-specific phospholipases; phospholipase Cβ and phospholipase Cγ (see chapter 3). Phospholipase Cβ is activated by seven transmembrane-spanning receptors coupled to heterotrimeric G proteins, while phospholipase Cγ is activated by tyrosine kinase coupled receptors. Two major classes of Ca^{2+} channels are present in the plasma membrane. Voltage-dependent channels, whose opening is triggered by the depolarization of the membrane and induces massive Ca^{2+} influx, are found in excitable cells such as neurons (Unwin 1989). A second class of voltage-independent Ca^{2+} channels which is yet poorly characterized is triggered by second messenger molecules.

INTEGRINS CAN ACTIVATE CA^{2+} TRANSIENTS

The demonstration that antibody-induced clustering of LFA-1 (αLβ2) on lymphocytes transiently raises intracellular calcium levels provided one of the first molecular studies

Signal Transduction by Integrins, by Paola Defilippi, Angela Gismondi, Angela Santoni and Guido Tarone. © 1997 Landes Bioscience.

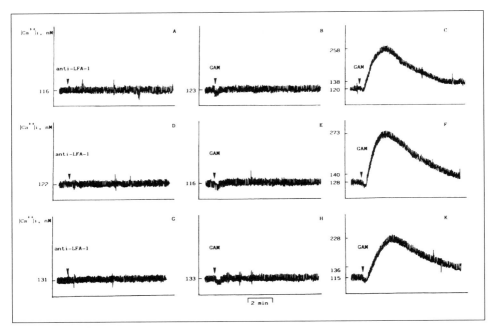

Fig. 4.1. LFA-1-induced Ca^{2+} response in purified lymphocyte subsets. Ca^{2+} response was evaluated in FURA-2 loaded CD4$^-$,CD16$^-$ (upper panel), CD8$^-$,CD16$^-$ (middle panel) and CD3$^-$,CD16$^-$ (lower panel) cells upon addition of 10 µg/ml anti-LFA-1 α chain antibody (A,D and G), 25 µg/ml polyclonal antimouse Ig (B, E and H), or 25 µg/ml polyclonal antimouse Ig after the cells had been treated with 10 µg/ml anti-LFA-1 α chain antibody for 20 min at 37°C (C,F and K). Results are representative of six separate experiments. With permission from Pardi R et al J Immunol 1989; 143:3157-3166. Copyright 1989 The American Association of Immunologists.

showing that integrins are capable of generating intracellular signals (Pardi et al 1989) (Fig. 4.1).

In addition to β2 integrins (Pardi et al 1989; Richter et al 1990; Jaconi et al 1991; Ng-Sikorski et al 1991; Altieri et al 1992; Eierman et al 1994; Hellberg et al 1994; Walzog et al 1994), also β1 (Leavesley et al 1993; Schwartz, 1993; Somogyi et al 1994; Palmieri et al 1995), β3 (Yamaguchi et al 1990; Miyauki et al 1991; Pelletier et al 1992; Leavesley et al. 1993; Paniccia et al 1993; Schwartz, 1993; Shankar et al 1993; Chenu et al 1994; Schwartz and Denninghoff, 1994; Sjaastad et al 1994; Zimolo et al 1994; Paniccia et al 1995) and β5 (Sjaastad et al 1994) integrin subunits were shown to increase cytoplasmic Ca^{2+} levels. The ability of integrin heterodimers to modify Ca^{2+} lev-

els depends both on the specific α subunit and on the cell type. Antibodies to β1, but not α2 or α5 integrin subunits trigger Ca^{2+} in endothelial cells (Schwartz, 1993; Leavesley et al 1993; Schwartz and Denninghoff,1994), indicating that integrin heterodimers other than α2β1 and α5β1 are responsible for β1-induced Ca^{2+} signaling in this cell type. On the other hand, α2β1 and α5β1 integrins induce a Ca^{2+} increase in acinar pancreatic (Somogyi et al 1994) and in natural killer cells, respectively (Palmieri et al 1995) (Fig. 4.2).

The rise in cytoplasmic Ca^{2+} levels occurs very rapidly after antibody-induced integrin clustering or binding of soluble ligands and peaks within 2 min (Pardi et al 1989; Paniccia et al 1993; Zimolo et al 1994) (Fig. 4.1). Slightly longer kinetics were ob-

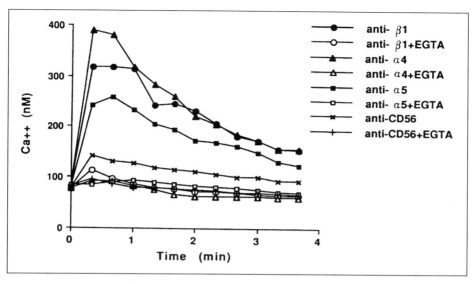

Fig. 4.2. α4β1 and α5β1 crosslinking by specific monoclonal antibodies induces Ca^{2+} mobilization in human natural killer (NK) cells. Highly purified in vitro cultured human NK cells were loaded with Fluo-3 and stimulated with different monoclonal antibodies crosslinked to polystyrene beads in the presence (open symbols) or absence (closed symbols) of 5 mM EGTA to chelate extracellular Ca^{2+}. Kinetic analysis of intracellular Ca^{2+} levels is reported. Antibodies to CD56 were used as negative control. Results of one representative experiment of three performed are shown. With permission from Palmieri G et al J Immunol. 1995;155:5314-5322. Copyright 1995 The American Association of Immunologists.

served in endothelial cells adhering on antibody or matrix-coated surfaces (Schwartz, 1993), which do not allow a rapid receptor-ligand interaction (Fig. 4.3). In some cases, interaction with a ligand-coated solid surface is required to trigger Ca^{2+} flux (Schwartz 1993; Palmieri et al 1995), while antibody-induced integrin clustering or soluble ligands are sufficient to promote increase in intracellular Ca^{2+} in other systems (Pardi et al 1989; Ng-Sikorski et al 1991; Altieri et al 1992; Paniccia et al 1993; Zimolo et al 1994). These differences may depend on the cell type studied or on the specific integrin complex involved and suggest that different mechanisms may link integrins to intracellular Ca^{2+} regulation. Integrin triggering of the calcium response may require accessory membrane molecules; the 50 kDa integrin-associated protein IAP is required for integrin-regulated calcium flux in endothelial cells (Schwartz et al 1993).

Extracellular Ca^{2+} is responsible for the integrin-induced intracellular Ca^{2+} increase in some systems (Schwartz, 1993; Palmieri et al 1995) (Fig. 4.2), but in others, both intracellular and extracellular stores contribute to the Ca^{2+} response (Pardi et al 1989; Ng-Sikorski et al 1991; Altieri et al 1992; Paniccia et al 1993; Petersen et al 1993; Zimolo et al 1994). While the mechanisms leading to integrin-mediated calcium influx across the plasma membrane are still unknown, the release from the intracellular stores may occur via phosphoinositide hydrolysis. Antibody-induced clustering of LFA-1 on activated lymphocytes causes hydrolysis of $PtdIns(4,5)P_2$ and generation of $Ins(1,4,5)P_3$, suggesting that the latter may be responsible for Ca^{2+} release from endoplasmic reticulum (Pardi et al 1989). Integrin-activated protein tyrosine kinases have also been implicated in the pathway leading to Ca^{2+} release in HL60 granulocytic

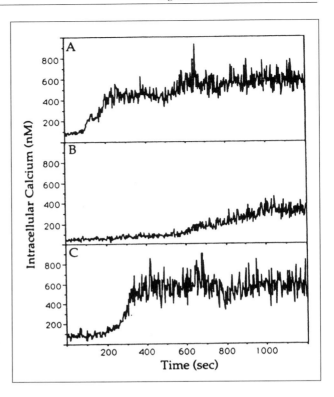

Fig. 4.3. Intracellular Ca²⁺ levels in human endothelial cells during spreading on vitronectin. Ca²⁺ levels as a function of time were recorded under microscope to detect changes in single cells. (A) Two cells in a cluster undergo abrupt increases at slightly different times; (B) one of the minor fraction of the cells (<20% of the population) where the increase in intracellular Ca²⁺ does not occur rapidly; (C) a cell undergoes a rapid increase in intracellular Ca²⁺. From Schwartz M et al J Cell Biol 1993; 120:1003-1010 by copyright permission of The Rockefeller University Press.

(Hellberg et al 1995) and epithelial (Somogyi et al 1994) cells. Their involvement, however, may vary considerably in different systems since inhibition of tyrosine kinases does not affect Ca²⁺ signaling in neutrophils (Walzog et al 1994). Moreover, in other cell types, tyrosine kinases were shown to be downstream of Ca²⁺ (Pelletier et al 1992).

What are the possible functional consequences of integrin-mediated Ca²⁺ signaling? In monocytes, this response is coupled with a transient activation state of the αMβ2 integrin and an increased avidity for its ligand fibrinogen (Altieri 1992). In neutrophils (Marks et al 1991; Damaurex et al 1996) and in endothelial cells, Ca²⁺ transients are important for cell spreading (Leavesley et al 1993; Schwartz, 1993) and migration, while in MDCK epithelial cells, Ca²⁺ response leads to enhanced adhesiveness (Sjaastad et al 1994). Thus Ca²⁺ may regulate the strength of the adhesive interaction via modulation of integrin affinity/

avidity for the ligand or cytoskeletal proteins. A cytoplasmic calcium-binding protein, calreticulin, is known to bind to the Gly-Phe-Phe-Lys-Arg highly conserved cytoplasmic sequence of integrin α chains (Rojiani et al 1991). Antisense oligonucleotides to calreticulin mRNA inhibit cell spreading and attachment, suggesting that this protein, by binding calcium, may regulate integrin functional state from the cytoplasmic side (Leung-Hagesteijn et al 1994; Coppolino et al 1995).

Ca²⁺ signaling via integrin receptors can be affected by extracellular stimuli. Calcitonin, a hormone that regulates Ca²⁺ levels in blood, downregulates integrin-induced Ca²⁺ flux in osteoclasts (Paniccia et al 1995). The chemotactic peptide f-Met-Leu-Phe impairs the Ca²⁺ signal capacity of β2 integrins in human neutrophils (Eierman et al 1994). Phosphorylation of Ser/Thr residues of the β2 integrin may regulate this function as suggested by the finding that phorbol myristate acetate-induced β2 phos-

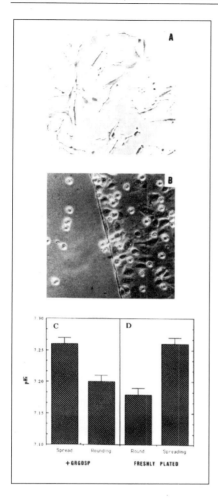

Fig. 4.4. Intracellular pH in cells as function of their shape during adhesion. (A) Phase-contrast micrograph of Balb/c 3T3 cells on plastic dishes 40 min after addition of 0.5 mM Gly-Arg-Gly-Asp-Ser-Pro peptide. (B) Phase-contrast micrograph of cells 45 min after plating on substrata with areas coated of fibronectin or of the nonadhesive polymer polyhydroxyethylmethacrylate (poly-HEMA). Cells attach and spread on fibronectin, while remaining completely round on the poly-HEMA. (C) Effect of rounding on intracellular pH (pHi). Cells on tissue culture plastic were loaded with 2',7' bis (2 carboxyethyl)-5,6 carboxyfluorescein (BCECF) and then treated with 0.5 mM Gly-Arg-Gly-Asp-Ser-Pro (+GRGDSP) to induce partial cell detachment from the substratum. Between 6 and 30 min after addition of the peptide pH was determined for 19 round and 18 flat cells. Graph shows means SEM. (D) Effect of spreading on intracellular pH. Freshly trypsinized cells were loaded with BCECF and plated on dishes with alternating regions of fibronectin and poly-HEMA. Between 15 and 45 min after plating pH was measured for 21 round cells on poly-HEMA and 19 spreading cells on fibronectin. With permission from Schwartz M et al Proc Natl Acad Sci USA 1989; 86:4525-4529.

phorylation abrogates β2-mediated Ca^{2+} flux induced via this integrin in HL60 granulocytic cells (Hellberg et al 1995).

INTRACELLULAR pH REGULATION BY INTEGRINS

The intracellular pH (pHi) increases of 0.2 units in response to several growth stimuli. The internal cellular pH (pHi) is controlled by the Na^+/H^+ exchange antiport system at the cell membrane whose function in response to extracellular stimuli is regulated by phosphorylation of its cytoplasmic domain in a protein kinase C-dependent manner (Wakabayashi et al 1992).

Cell matrix interactions via integrin receptors can induce cytoplasmic alkaliniza-

tion. Initial observations indicated that adhesion of fibroblasts to a fibronectin-coated surface induces an increase in pHi of about 0.15 pH units as compared with round cells (Schwartz et al 1989; Schwartz et al 1990). The pHi increased proportionally to the degree of cell spreading on the substratum (Fig. 4.4) (Schwartz et al 1989; Ingberg et al 1990; Galkina et al 1992) presumably as a consequence of the progressive engagement of an increasing number of integrin receptors. Subsequently several integrin heterodimers, including $\alpha5\beta1$, $\alpha\nu\beta3$ and $\alpha L\beta2$ as well as different matrix proteins, have been shown to promote increased pHi (Schwartz et al 1991a and 1991b), indicating that this signaling pathway is shared by many

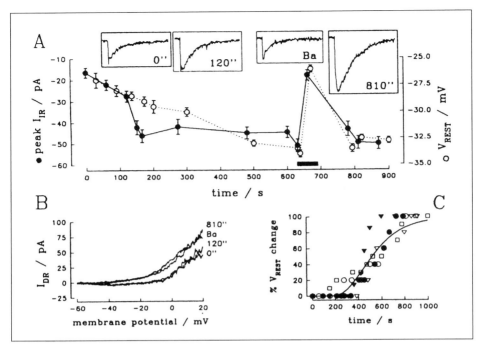

Fig. 4.5. (A) Time course of the peak K_{IR} currents (peak I_{IR} left axis) and V_{rest} (right axis) after mouse N1 neuroblastoma cells contact to a fibronectin-coated bead. Cells were plated on bovine serum albumin-coated dish containing serum free DME to which fibronectin-coated beads (diameter 11,6 μm) were added and left to sediment. A single cell was then patch-clamped (whole cell) and moved in contact with a coated bead. V_{rest} and K_{IR} measurements were taken in current- or voltage-clamp conditions, respectively. (Insets) K_{IR} current recordings elicited at -100 mV (for 250 ms) from a holding potential of -10 mV at the indicated times. After 650 s from the initial contact and for the indicated time (black horizontal bar) 2 mM Ba^{2+} was applied and then washed out. Note the blocking effect of Ba^{2+} on the K_{IR} current and the concomitant depolarization. (B) I-V plot of the K_{DR} current taken at the indicated times after contact in the same cell used in A, by applying a ramp protocol between -80 and +20 mV. Note the lack of significant activation of this current after contact and its insensitivity to Ba^{2+}. Recordings were corrected for a linear leakage (leakage conductance 1.9/4.5 GΩ). (C) Time course of the percent (%) V_{rest} change (hyperpolarization) recorded in five experiments similar to that shown in A. Data were normalized to the maximal hyperpolarization obtained in each experiment. Different symbols refer to different cells. The best fitting line corresponds to the function $(1-e^{-(t-91)/192})$. From Arcangeli A et al J Cell Biol 1993;122:1131-1143 by copyright permission of The Rockefeller University Press.

members of the integrin family. Increase in intracellular pH can be uncoupled from changes in cell shape induced by adhesion to a matrix substratum since soluble divalent antibodies to β1 can trigger increased pHi in round endothelial cells (Schwartz et al 1991b).

Interestingly, the fibronectin receptor and bFGF both increase pHi in capillary endothelium by activating the Na^+/H^+ antiporter, but apparently via separate pathways (Ingberg et al 1990).

The processes by which integrin engagement leads to activation of the Na^+/H^+ antiporter are still unknown. Morphological data indicate that the ubiquitous isoform of the antiporter, NHE-1, colocalizes at cell substratum contact sites with vinculin, talin and actin, thus suggesting a close association with integrin-containing adhesion structures (Grinstein et al 1993).

INTEGRINS CAN REGULATE MEMBRANE POTENTIAL

Fibronectin and laminin and their surface receptors can induce membrane hyperpolarization in both erythroleukemia and neuroblastoma cells (Arcangeli et al 1991; Becchetti et al 1992; Arcangeli et al 1993). These measurements, carried out with patch-clamp techniques in single cells, indicate that a 15-20 mV hyperpolarization occurs within 5-6 min and is completed about 10 min after initial contact with fibronectin-coated beads (Fig. 4.5). The integrin-induced hyperpolarization is due to the activation of Ca^{2+}-dependent K^+ channels and parallels cell spreading (Arcangeli et al 1991; Becchetti et al 1992).

In murine neuroblastoma cells hyperpolarization occurred through a G-protein-mediated activation of an "inward rectifier" K^+ channel. This integrin-mediated response is absolutely required for neurite elongation on fibronectin as a neuroblastoma clone lacking this channel does not hyperpolarize nor elongate neurites in response to integrin occupancy (Arcangeli et al 1993).

References

Altieri DC, Stamnes SJ, Gahmberg CG (1992) Regulated Ca^{2+} signaling through leukocyte CD11b/CD18 integrin. Biochem J 288:465-473.

Arcangeli A, Becchetti A, Del Bene MR, Wanke E, Olivotto M (1991) Fibronectin-integrin binding promotes hyperpolarization of murine erythroleukemia cells. Biochem Biophys Res Commun 177:1266-1272.

Arcangeli A, Becchetti A, Mannini A, Mugnai G, Defilippi P, Tarone G, Del Bene MR, Barletta E, Wanke E, Olivotto M (1993) Integrin-mediated neurite outgrowth in neuroblastoma cells depends on the activation of potassium channels. J Cell Biol 122:1131-1143.

Becchetti A, Arcangeli A, Del Bene MR, Olivotto M, Wanke E (1992) Response to fibronectin-integrin interaction in leukaemia cells: delayed enhancing of a K^+ current. Proc R Soc Lond B Biol Sci 248:235-240.

Berridge MJ (1993) Inositol trisphosphate and calcium signalling. Nature 361:315-325.

Chenu C, Colucci S, Grano M, Zigrino P, Barattolo R, Zambonin G, Baldini N, Vergnaud P, Delmas PD, Zallone AZ (1994) Osteocalcin induces chemotaxis, secretion of matrix proteins, and calcium-mediated intracellular signaling in human osteoclast-like cells. J Cell Biol 127:1149-1158.

Clapham DE (1995) Calcium signalling. Cell 80:259-268.

Coppolino M, Leung-Hagesteijn C, Dedhar S, Wilkins J (1995) Inducible interaction of integrin $\alpha2\beta1$ with calreticulin. Dependence on the activation state of the integrin. J Biol Chem 270:23132-23138.

Damaurex N, Downey GP, Waddell TK, Grinstein S (1996) Intracellular pH regulation during spreading of human neutrophils. J Cell Biol 133:1391-1402.

Eierman D, Hellberg C, Sjolander A, Andersson T (1994) Chemotactic factor receptor activation transiently impairs the Ca^{2+} signaling capacity of $\beta2$ integrins on human neutrophils. Exp Cell Res 215:90-96.

Galkina SI, Sud'Ina GF, Margolis LB (1992) Cell-cell contacts alter intracellular pH. Exp Cell Res 200:211-214.

Grinstein S, Woodside M, Waddell TK, Downey GP, Orlowski J, Pouyssegur J, Wong DC, Foskett JK (1993) Focal localization of the NHE-1 isoform of the Na^+/H^+ antiport: assessment of effects on intracellular pH. EMBO J 12:5209-5218.

Hellberg C, Eierman D, Sjolander A, Andersson T (1995) The Ca^{2+} signaling capacity of the $\beta2$ integrin on HL60-granulocytic cells is abrogated following phosphorylation of its CD18 chain: relation to impaired protein tyrosine phosphorylation. Exp Cell Res 217:140-148.

Ingber DE, Prusty D, Frangioni JV, Cragoe EJ Jr, Lechene C, Schwartz MA (1990) Control of intracellular pH and growth by fibronectin in capillary endothelial cells. J Cell Biol 110:1803-1811.

Jaconi ME, Theler JM, Schlegel W, Appel RD, Wright SD, Lew PD (1991) Multiple elevations of cytosolic-free Ca^{2+} in human neutrophils: initiation by adherence receptors of the integrin family. J Cell Biol 112:1249-1257.

Leavesley DI, Schwartz MA, Rosenfeld M, Cheresh DA (1993) Integrin β1- and β3-mediated endothelial cell migration is triggered through distinct signaling mechanisms. J Cell Biol 121:163-170.

Leung-Hagesteijn CY, Milankov K, Michalak M, Wilkins J, Dedhar S (1994) Cell attachment to extracellular matrix substrates is inhibited upon downregulation of expression of calreticulin, an intracellular integrin α-subunit-binding protein. J Cell Sci 107:589-600.

Marks PW, Hendey B, Maxfield FR (1991) Attachment to fibronectin or vitronectin makes human neutrophil migration sensitive to alterations of cytosolic free calcium concentration. J Cell Biol 112:149-158.

Miyauki A, Alvarez J, Greenfield E, Teti A, Grano M, Colucci S, Zambonin-Zallone A, Ross FP, Teitelbaum SL, Cheresh D, Hruska KA (1991) Recognition of osteopontin and related peptides by an αV/β3 integrin stimulate cell signals in osteoclasts. J Biol Chem 266:20396-20374.

Ng-Sikorski J, Andersson R, Patarroyo M, Andersson T (1991) Calcium signaling capacity of the CD11b/CD18 integrin on human neutrophils. Exp Cell Res 195:504-508.

Palmieri G, Serra A, De Maria R, Gismondi A, Milella M, Piccoli M, Frati L, Santoni A (1995) Crosslinking of α4β1 and α5β1 fibronectin receptors enhances natural killer cell cytotoxic activity. J Immunol 155:5314-5322.

Paniccia R, Colucci S, Grano M, Serra M, Zallone AZ, Teti A (1993) Immediate cell signal by bone-related peptides in human osteoclast-like cells. Am J Physiol 265:C1289-1297.

Paniccia R, Riccioni T, Zani BM, Zigrino P, Scotlandi K, Teti A (1995) Calcitonin downregulates immediate cell signals induced in human osteoclast-like cells by the bone sialoprotein-IIA fragment through a postintegrin receptor mechanism. Endocrinology 136:1177-1186.

Pardi R, Bender JR, Dettori C, Giannazza E, Engelman EG (1989) Heterogeneous distribution and transmembrane signaling properties of lymphocyte function-associated antigen (LFA-1) in human lymphocyte subsets. J Immunol 143:3157-3166.

Pelletier AJ, Bodary SC, Levinson AD (1992) Signal transduction by the platelet integrin αIIbβ3: induction of calcium oscillations required for protein-tyrosine phosphorylation and ligand-induced spreading of stably transfected cells. Mol Biol Cell 3:989-998.

Petersen M, Williams JD, Hallett MB (1993) Cross linking of CD11b or CD18 signals release of Ca^{2+} from intracellular stores in neutrophils. Immunology 80:157-159.

Richter J, Ng-Sikorski J, Olsson I, Andersson T (1990) Tumor necrosis factor-induced degranulation in adherent human neutrophils is dependent on CD11b/CD18-integrin-triggered oscillations of cytosolic free Ca^{2+}. Proc Natl Acad Sci USA 87:9472-9476.

Rojiani MV, Finlay BB, Gray V, Dedhar S (1991) In vitro interaction of a polypeptide homologous to human Ro/SS-A antigen (calreticulin) with a highly conserved amino acid sequence in the cytoplasmic domain of integrin α subunits. Biochemistry 30:9859-9866.

Schatzmann HJ (1989) The calcium pump of the surface membrane and of the sarcoplasmic reticulum. Ann Rev Physiol 51:473-489.

Schwartz MA, Both G, Lechene C (1989) Effect of cell spreading on cytoplasmic pH in normal and transformed fibroblasts. Proc Natl Acad Sci USA 86:4525-4529.

Schwartz MA, Brown EJ, Fazeli B (1993) A 50 kDa integrin associated protein is required for integrin-regulated calcium entry in endothelial cells. J Biol. Chem 268:19931-19934.

Schwartz MA, Cragoe EJ Jr, Lechene CP (1990) pH regulation in spread cells and round cells. J Biol Chem 265:1327-1332.

Schwartz MA, Denninghoff K (1994) αv integrins mediate the rise in intracellular calcium in endothelial cells on fibronectin even though they play a minor role in adhesion. J Biol Chem 269:11133-111337.

Schwartz MA, Ingber DE, Lawrence M, Springer TA. Lechene C (1991a) Multiple integrins share the ability to induce elevation of intracellular pH. Exp Cell Res 195:533-535.

Schwartz MA, Lechene C, Ingber DE (1991b) Insoluble fibronectin activates the Na/H antiporter by clustering and immobilizing integrin $\alpha5/\beta1$, independent of cell shape. Proc Natl Acad Sci USA 88:7849-7853.

Schwartz MA (1993) Spreading of human endothelial cells on fibronectin or vitronectin triggers elevation of intracellular free calcium. J Cell Biol 120:1003-1010.

Shankar G, Davison I, Helfrich MH, Mason WT, Horton MA (1993) Integrin receptor-mediated mobilization of intranuclear calcium in rat osteoclasts. J Cell Sci 105:61-68.

Sjaastad MD, Angres B, Lewis RS, Nelson WJ (1994) Feedback regulation of cell-substratum adhesion by integrin-mediated intracellular Ca^{2+} signaling. Proc Natl Acad Sci USA 91:8214-8218.

Somogyi L, Lasic Z, Vukicevic S, Banfic H (1994) Collagen type IV stimulates an increase in intracellular Ca^{2+} in pancreatic acinar cells via activation of phospholipase C. Biochem J 299:603-611.

Unwin N (1989) The structure of ion channels in membranes of excitable cells. Neuron 3:665-676.

Wakabayashi S, Sardet C, Fafournoux P, Counillon L, Meloche S, Pages G, Pouyssegur J (1992) Structure function of the growth factor-activable Na^+/H^+ exchanger (NHE1). Rev Physiol Biochem Pharmacol 119:157-86.

Walzog B, Seifert R, Zakrzewicz A, Gaehtgens P, Ley K (1994) Crosslinking of CD18 in human neutrophils induces an increase of intracellular free Ca^{2+}, exocytosis of azurophilic granules, quantitative upregulation of CD18, shedding of L-selectin, and actin polymerization. J Leukoc Biol 56:625-635.

Yamaguchi A, Tanoue K, Yamazaki H (1990) Secondary signals mediated by GPIIb/IIIa in thrombin-activated platelets. Biochim Biophys Acta 1054:8-13.

Zimolo Z, Wesolowski G, Tanaka H, Hyman JL, Hoyer JR, Rodan GA (1994) Soluble $\alpha v\beta3$-integrin ligands raise $[Ca^{2+}]i$ in rat osteoclasts and mouse-derived osteoclast-like cells. Am J Physiol 266:C376-381.

CHAPTER 5

Activation of MAPK Pathways

PATHWAYS CONTROLLING MAPK ACTIVATION

Mitogen-activated protein kinases (MAPKs) comprise a group of protein serine/threonine kinases which are important intermediates in signal transduction pathways initiated by many types of cell surface receptors. Regulation of MAPK activity has been implicated in a wide array of physiological processes such as cell proliferation, differentiation, development and survival.

At least three distinct groups of MAPKs have been identified in mammals, including extracellular signal-regulated kinases (Erk1 and Erk2), Jun N-terminal kinase (Jnk) and p38. They are activated in response to distinct extracellular stimuli and have different substrate specificity, although some crosstalk between their activation pathways exists (Davis, 1994; Seger and Krebs, 1995; Su and Karin, 1996).

A common characteristic in MAPK activation is the presence of a protein kinase cascade leading to activation of a MAPKK kinase (MAPKKK) that activates MAPK kinase (MAPKK) that in turn activates MAPK by dual phosphorylation at conserved threonine (Thr) and tyrosine (Tyr) residues (Fig. 5.1).

The first MAPKs identified in mammals, Erk1 and Erk2 (p44 and p42 MAPK, respectively) show a 90% identity. Erk1 and Erk2 activity is rapidly stimulated by receptors endowed with tyrosine kinase activity, connected to nonreceptor tyrosine kinase, or coupled to G proteins. Raf-1 is the most extensively studied MAPKKK involved in Erk activation. Raf-1 kinase activity is dependent on activation of p21Ras protein. The interaction of Ras with Raf, however, is only responsible for Raf recruitment to the plasma membrane where it is phosphorylated and activated by specific kinase(s) whose identity is still unknown; PKC and PTK have been suggested as possible candidates (McCormick, 1994).

p21Ras protein is a small membrane-associated GTPase whose activation requires conversion from the inactive GDP-bound to the active GTP-bound state (Fig. 5.2). The ratio of the two forms is regulated by the opposing effects of guanine nucleotide exchange factors (GEFs), which favor the substitution of GDP by a free GTP nucleotide, and GTPase activating enzymes (GAPs), which catalyse the intrinsic rate of Ras GTP hydrolysis. One of the best characterized GEF is Sos which is recruited to the plasma membrane by the adaptor protein Grb2 (McCormick, 1993; Schlessinger, 1993). Grb2 is coupled to phosphotyrosine receptors either directly or via the SH2-containing protein Shc. Shc is a modular molecule char-

Signal Transduction by Integrins, by Paola Defilippi, Angela Gismondi, Angela Santoni and Guido Tarone. © 1997 Landes Bioscience.

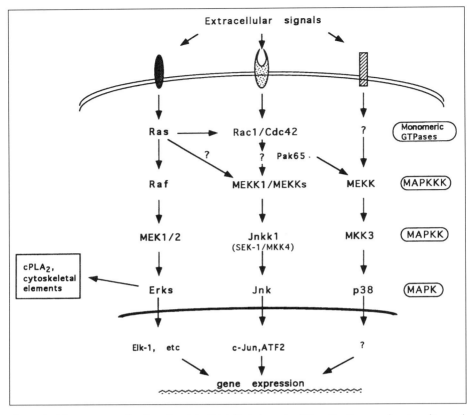

Fig. 5.1. Kinase cascades involved in Erk1/2, Jnk and p38 activation and cytosolic and nuclear targets.

acterized by a carboxyterminal SH2 domain, a central collagen-homology domain containing a Tyr residue that upon phosphorylation become a Grb2-binding site, and an amino terminal phosphotyrosine-binding domain (PTB) which is different from that of SH2. Shc functions as an alternative docking site for the Grb2-Sos complex in the Ras signaling pathway (Bonfini et al 1996).

A large number of proteins located both in the cytoplasm and the nucleus can be phosphorylated by Erks at Ser/Thr-Pro motif. Substrates of Erks include kinases, lipases, transcription factors and cytoskeletal proteins. The two best characterized Erk substrates are the cytoplasmic phospholipase A_2 (cPLA$_2$) and the transcription factor Elk-1. Other transcription factors and nuclear proteins which can be substrates for

Erks are c-Jun, which can also be a substrate of Jnk, c-fos, ATF-2, c-Myc, c-Myb Ets2, NF-IL6, TAL-1 p53 and RNA polymerase II. Cytoskeletal elements known as Erk substrates include microtubule-associated proteins (MAP-1, MAP-2, MAP-4), Tau, etc. Their phosphorylation may play an important role in the regulation of cytoskeletal rearrangements and cell shape (Davis, 1994; Seger and Krebs, 1995; Su and Karin, 1996).

Similarly to Erks, Jnk is activated by phosphorylation on conserved threonine and tyrosine residues by the MAPKK, SEK-1 (MKK4 or JNKK1) which in turn is phosphorylated and activated by the MAPKK kinase (MAPKKK), MEKK1 (Fig. 5.1). Jnk activation has been observed following treatment of cells with UV radiation, environmental stress, inflammatory cytokines,

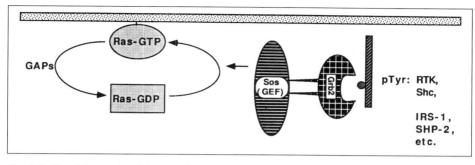

Fig. 5.2. Possible mechanisms involved in p21 Ras activation.

including tumor necrosis factor (TNF) and interleukin-1 (IL-1), and growth factors (Davis, 1994; Seger and Krebs, 1995; Su and Karin, 1996). The events involved in the activation of Jnk pathway by these stimuli are not yet fully clarified. Unlike Erks, controversial evidence is available on the regulation of Jnk activation by Ras. Other small G proteins, belonging to the Rho family such as Rac and Cdc42, but not Rho, are more effective activators of Jnk than Erk1 and Erk2 (Vojtek and Cooper, 1995). Activated Cdc42 and Rac can bind and in turn activate a STE20 related protein kinase named Pak-1 which acts upstream of MEKK1. However, it is still unknown whether Pak-1 or other members of the Pak family play a role in the activation of Jnk.

The major Jnk substrate is c-Jun, although Jnk activity is also directed toward other transcription factors, including Elk-1 and ATF-2 (Davis, 1994; Su and Karin, 1996).

p38 MAPK pathway is activated in response to LPS and osmotic shock, as well as by the same stimuli that activate Jnks (Davis, 1994; Su and Karin, 1996). p38 is specifically activated by the MAPKK, MKK3, which does not affect Jnk and Erk activity (Fig. 5.1); in addition, it is also target of the Jnk kinase pathway, MAPKK, MKK4 (JNKK1). Additional components of the p38 MAPK pathway have not been identified. However, the Ras family GTPases, Rac1 and Cdc42 and Pak-1, Pak-3 and GC kinase have been implicated also in the control of p38 MAPK signaling pathway.

The cellular substrates of p38 MAPK pathway are largely unknown; its involvement in the phosphorylation and activation of ATF-2 has been reported (Su and Karin, 1996).

INTEGRIN-STIMULATED ERK1/ERK2 ACTIVATION

A role for Ras/MAPK pathways in coupling integrins to gene expression and control of cell proliferation has been recently proposed. In this chapter we will discuss the mechanisms of integrin-mediated Ras/MAPK activation, while the downstream effects on gene expression and proliferation are discussed in chapters 7 and 8.

The first indication that cell adhesion to extracellular matrix components activates MAPK was reported by Chen and coworkers (Fig. 5.3) (Chen et al 1994). Their data demonstrate that adhesion of Swiss 3T3 or REF52 fibroblast cell lines to fibronectin, laminin or Arg-Gly-Asp-containing peptides leads to a strong and prompt activation of both p42 and p44 forms of Erk. Moreover, following adhesion of rat fibroblasts to fibronectin, Erk translocation from the cytoplasm to the nucleus parallels enzyme activation. They also provide evidence that integrity of the actin cytoskeleton is required, in that, treatment with cytochalasin D, which disrupts actin microfilaments, almost completely blocks adhesion-induced Erk activation.

Activation of Erk-1 and Erk-2 MAPKs following long-term adhesion of G_0-synchronized NIH-3T3 cells to fibronectin and

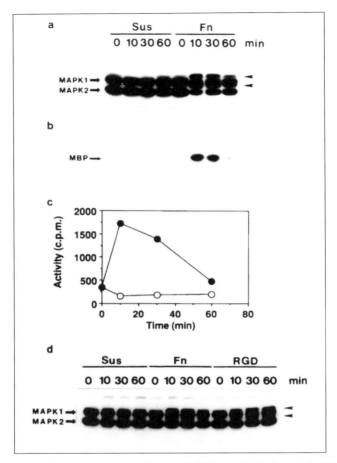

Fig 5.3. Adhesion to fibronectin or Arg-Gly-Asp (RGD) peptide induces MAPK (Erk) activation. In panels a-c, Swiss 3T3 cells were grown to confluence in complete medium, disso-
ciated with trypsin/EDTA and then kept in suspension in serum-free medium (Sus) or plated
on to a substratum coated with 10 μg/ml fibronectin (Fn) and incubated at 37°C for 0, 10,
30 and 60 min. Following the incubations, total cell lysates were prepared and MAPK
activity was examined by mobility shift assay and by immunocomplex kinase assay.
(Throughout the legend and on the figure, the 44 kDa form of MAPK is designated MAPK1,
while the 42 kDa form is designated MAPK2). Panel a shows the mobility shift of both
MAPK1 and MAPK2; the arrowheads indicate slower-migrating forms of MAPK1 and MAPK2
corresponding to the activated forms. In panel b, the activity of MAPKs in immunoprecipitates was measured after addition of MBP (myelin basic protein) as a substrate; phosphorylated MBP was separated by a 15% SDS gel and autoradiographed. The MBP bands were
also excised and counted in a scintillation counter (panel c). The filled circles represent
radiophosphate incorporeted into MBP in cells plated on fibronectin, while the open circles
represent incorporation in control cells. In panel d, confluent Swiss 3T3 cells were kept in
suspension (Sus) or plated on substrata coated with 10 μg/ml fibronectin (Fn) or 100 μM
synthetic Arg-Gly-Asp-Ser-Pro-Lys peptide (RGD) and incubated for varying times at 37°C.
After the incubations, MAPK activity was evaluated by the mobility shift assay as described
above. The arrowheads indicate the activated forms of MAPK1 and MAPK2. With permission from Chen Q et al J Biol Chem 1994; 269:26602-26605.

their association with integrin-dependent shape changes has been described (Zhu and Assoian, 1995). These authors compared the ability of growth factors and extracellular matrix components to activate Erk. Their results indicate major differences in the ability of these two stimuli to activate Erk. Growth factor-dependent activation of Erk is rapid and transient (maximal at 10 min), whereas the activation in response to fibronectin is gradual, persistent (maximal at 1 h and remained elevated for at least 3 h), and is associated with cell spreading but not with cell attachment. Activation of Erk also occurs following cell adhesion to vitronectin and type IV collagen with a kinetics that correlates with cell spreading to these extracellular matrix components. These results suggest that combined stimulation of Erk activity by integrins and growth factor receptors may be required for optimal cell cycle progression.

These data only partially confirm studies by Chen et al 1994 which report a rapid and transient Erk stimulation by fibronectin. The basis for these differences is unclear, but they may result from the different cell lines used or that G_0-arrested (Zhu and Assoian 1995) and asynchronous (Chen et al 1994) cell populations were used in the two experiments.

Direct involvement of β1 integrins in Erk1 and Erk2 MAPK activation has been reported (Morino et al 1995). The data show that both Erk1 and Erk2 are rapidly phosphorylated upon antibody-mediated crosslinking of β1 integrins or adhesion of human skin fibroblasts to fibronectin.

RAS INVOLVEMENT IN INTEGRIN-DEPENDENT ERK ACTIVATION

Existing evidence suggests a role for p21 Ras in the upstream events leading to integrin-induced Erk activation.

Induction of p21 Ras activity has been observed in Jurkat T cells following ligation of α2β1 integrin by collagen or anti-α2 and anti-β1 stimulatory antibodies (Kapron-Bras et al 1993).

Activation of p21Ras has been also observed in adherent human neutrophils following antibody- or ICAM-1-induced engagement of β2 integrins. Ras activation correlated with tyrosine phosphorylation of the protooncogene product Vav, an hematopoietic-specific guanine nucleotide exchange factor (GEF) reported to activate Ras in T lymphocytes (Zheng et al; 1996; Gulbin, 1993) as well as Rac (Bustelo et al 1994; Khosravi-Far et al; 1994; Crespo et al 1997).

Moreover, a role for Ras in the fibronectin-dependent activation of Erk2 in adherent NIH3T3 cells has been demonstrated by the expression of a dominant-inhibitory Ras mutant. The Ras dominant-inhibitory mutant inhibits Erk activity, but not cell adhesion, spreading or focal contact, and stress fiber formation, suggesting that integrin-mediated signals involved in regulating cell morphology diverge from those regulating Erk activation at a level upstream of Ras activation (Clark and Hynes, 1996).

The Shc-Grb2 complex may play a role in the mechanisms involved in the recruitment of Ras to the plasma membrane following integrin engagement. Tyrosine phosphorylation of Shc and its association with either the phosphorylated β3 subunit of GPIIbIIIa (αIIbβ3) and Grb2 has been reported following thrombin-induced platelet aggregation (Law et al 1996).

Binding of the adaptor protein Shc to the tyrosine phosphorylated integrin β4 subunit, has been also observed in primary human keratinocytes following α6β4 ligation. Upon this interaction, Shc undergoes tyrosine phosphorylation and combines with Grb2 (Mainiero et al 1995).

A direct demonstration of the involvement of the Ras-Erk pathway in the signaling of some β1 and αvβ3 integrins has been recently provided (Fig. 5.4) (Wary et al 1996). The results from this study indicate the existence of two classes of integrins differing in their ability to activate the Ras-Erk pathway via Shc when engaged by specific ligands or antibodies. Tyrosine phosphorylation of Shc and its association with Grb2 has been observed following stimulation via

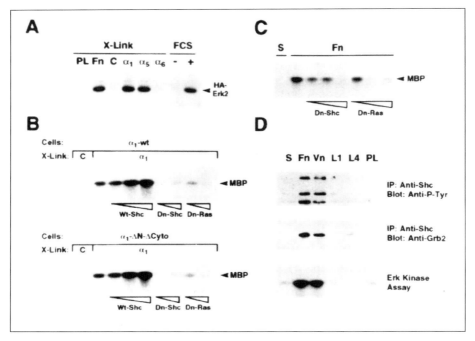

Fig. 5.4. Role of Shc in activation of MAPK by integrins. (A) NIH-3T3 cells expressing a single chain tail-less α1 subunit (NIH-3T3-α1-ΔN-ΔCyto) were transiently transfected with hemoagglutinin (HA)-tagged Erk2 plasmid, then growth factor-starved and stimulated in suspension with beads coated with poly-L-lysine (PL), fibronectin (Fn) or Mabs W6.32 to major histocompatibility class I antigen (MHCI)(C), TS2/7 (α1), 5H10-27 (α5) and GoH3 (α6). As a control, the cells were stimulated with 10% FCS (plus) or left untreated (minus). HA-Erk2 was immunoprecipitated and subjected to in vitro kinase assay.
(B) NIH-3T3 cells expressing α1 wild-type (NIH-3T3-α1-WT) and NIH-3T3-α1-ΔN-ΔCyto cells were transiently transfected with 3 µg of Erk2-HA plasmid alone or in combination with 2.5, 5.0 and 10 µg of wild-type Shc plasmid (WT-Shc), 2.5 and 5.0 µg of dominant negative Shc plasmid (Dn-Shc), and 2.5 and 5.0 µg of dominant-negative Ras plasmid (DN-Ras). The cells were stimulated with Mabs W6.32 (c) or TS2/7 (α1). Immunoprecipitated HA-Erk2 was subjected to in vitro kinase assay with MBP (myelin basic protein) as a substrate.
(C) NIH 3T3 cells were transiently transfected with 3 µg of HA-Erk2 plasmid alone or in combination with 2.5, 5.0 and 10 µg of dominant-negative Shc (Dn-Shc) or Ras plasmid (DN-Ras). After starvation, the cells were either kept in suspension or plated on fibronectin-coated dishes. HA-Erk2 was immunoprecipitated and subjected to in vitro kinase assay with MBP as a substrate.
(D) Human umbilical vein endothelial cells (HUVECs) were growth factor-starved and either kept in suspension or plated on dishes coated with poly-L-lysine (PL), fibronectin (Fn), vitronectin (Vn), laminin 1 (L1) or laminin 4 (L4). Lysates were immunoprecipitated with anti-Shc Mab clone 8 followed by immunoblotting with anti-pTyr Mab RC-20 (top) or anti-Grb2 serum (middle). HUVECs were transiently transfected with HA-Erk2 plasmid prior to plating on the various substrata. HA-Erk2 was immunoprecipitated and subjected to in vitro kinase assay with MBP as a substrate (bottom). With permission from Wary KK et al Cell 1996; 87:733-743. Copyright Cell Press.

α5β1, α1β1, αvβ3, but not α2β1, α3β1 and α6β1. The association of Shc with integrins is specified by sequences contained in the membrane-proximal portion of the extracellular domain and/or the transmembrane segment of the α subunit and is independent from the β subunit. This association is indirect and is likely to be mediated by caveolin, a transmembrane protein implicated in linking a variety of cell surface receptors that lack a cytoplasmic domain to the intracellular signaling pathway. Caveolin is constitutively associated with β1 integrins and forms a complex with Shc upon stimulation with anti-β1 antibody. Using NIH-3T3 cell transfectants expressing a single-chain tail-less α1 subunit, Wary and coworkers also show that, unlike the wild-type α1β1 integrin which can both recruit Shc and activate p125Fak, this mutant can recruit Shc, but not activate p125Fak; on the other hand wild-type α6β1 integrin activates p125Fak, but does not recruit Shc. Thus, the recruitment and tyrosine phosphorylation of Shc and the activation of p125Fak are separate events. Phosphorylated Shc associates with β1 and β4 integrins by different mechanisms: indeed, for β1 integrins, this association seems to be mediated by caveolin, whereas a direct interaction between tyrosine phosphorylated Shc and tyrosine phosphorylated β4 subunit is suggested (Wary et al 1996; Mainiero et al 1995).

An association between β1 or β4 integrins and Shc-Grb2-Sos complex in the activation of Ras-MAPK cascade has been demonstrated (Wary et al 1996; Mainiero et al 1997). Indeed, stimulation of primary human keratinocytes through β1 integrins resulted in activation of p21Ras; the dominant negative of Shc and Ras prevented the activation of Erk2 in response to cell adhesion to fibronectin. Moreover, only integrins that activate Shc were found to induce Erk activation, whereas the integrins that activate p125Fak, but do not recruit Shc, were ineffective in Erk activation. Overall, these results implicate Shc and not p125Fak, in the activation of Ras-MAPK pathway by integrins (Fig. 5.5, panel A).

A role for p125Fak in the integrin-mediated activation of Ras-MAPK has been previously suggested (Fig. 5.5, panel B) (Schlaepfer et al 1994; Schlaepfer and Hunter 1996). These studies demonstrate that interaction of NIH3T3 fibroblasts with fibronectin induces tyrosine phosphorylation of p125Fak and promotes its association with the adaptor protein Grb2. In addition, formation of p125Fak-Grb2-Sos complex parallels activation of Erk kinases, but direct evidence of p125Fak involvement in Ras or Erk activation is not available. Thus, the role of p125Fak-Grb2 association in integrin-stimulated Ras-MAPK pathway requires further investigation.

RHO INVOLVEMENT IN INTEGRIN-MEDIATED ERK ACTIVATION

Although many observations indicate the involvement of Ras in the integrin-mediated activation of the Erk pathway, Ras independent signaling events have also been implicated in its control.

In this regard, it has been demonstrated that a dominant negative inhibitor of Ras-dependent signaling fails to block activation of Raf-1 and MEK induced upon fibronectin or antibody-mediated crosslinking of β1 integrins on mouse NIH3T3 fibroblasts. In addition, while treatment with the EGF clearly increased GTP-loading of Ras, little effect was observed in response to integrin-dependent cell adhesion (Chen et al 1996).

The involvement of Rho in the fibronectin-dependent activation of MAPK has been reported. A dominant negative Rho mutant or C3 exoenzyme inhibit MAPK activity induced following NIH 3T3 cell adhesion to fibronectin. Conversely, activation of Rho by expression of a constitutively active mutant or by the guanine nucleotide exchange protein Lbc enhanced the fibronectin-induced MAPK activation (Renshaw et al 1996).

The molecular mechanisms by which Rho can activate MAPK have not been elucidated. Since MAPK activation in response to adhesion requires cell spreading and actin cytoskeleton organization (Chen et al

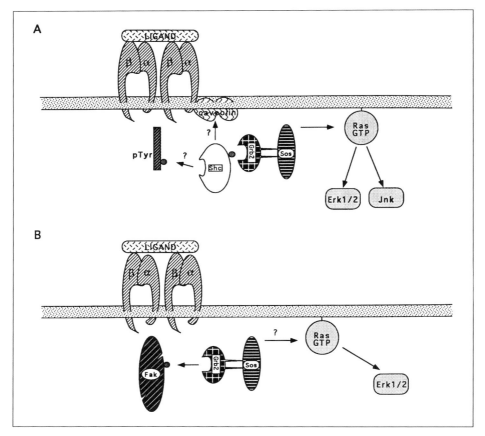

Fig. 5.5. Possible integrin-mediated signaling pathways leading to MAPK activation.

1994; Bohmer et al 1996), it is possible that Rho is indirectly controlling MAPK by favoring actin stress fiber organization (see also chapter 6).

Evidence also indicates that full activation of Erk following α6β4 ligation requires the activity of both Rho and Ras. In this system, Erk activation is inhibited by dominant negative Shc, Ras and RhoA, but not by dominant negative Cdc42 or Rac (Mainiero et al 1997).

JNK ACTIVATION BY INTEGRINS

Unlike Ras-Erk, very few data are available on the activation of Jnk and p38 MAPK following integrin stimulation.

It has been shown that ligation of α6β4 results in activation of Jnk kinase which is inhibited by the dominant negative of Ras

and Rac and by the PI-3 kinase inhibitor, wortmannin. These data suggest that PI-3 kinase, a downstream target-effector of Ras, is involved in α6β4-triggered Rac-Jnk signaling pathway by regulating Rac (Mainiero et al 1997).

Activation of Jnk has also been reported by Miyamoto and coworkers (Miyamoto et al 1995). Integrin ligation by fibronectin, multivalent Arg-Gly-Asp ligand, adhesion-blocking antibody or noninhibitory antibodies is sufficient to induce rapid activation of the Jnk pathway, in a manner temporally distinct from Erk activation. Interestingly, cytochalasin D inhibits activation of the Erk pathway, while it does not block activation of Jnk pathway. Thus, integrins can mediate the activation of two distinct MAPK pathways with different time courses and different responses to disruption of the actin cytoskeleton.

REFERENCES

Bohmer RM, Scharf E, Assoian RK (1996) Cytoskeletal integrity is required throughout the mitogen stimulation phase of the cell cycle and mediates the anchorage-dependent expression of cyclin D1. Mol Biol Cell 7:101-111.

Bonfini L, Migliaccio E, Pelicci G, Lanfrancone L, Pelicci PG (1996) Not all Shc's roads lead to Ras. Trends Biochem Sci 21:257-261.

Bustelo XR, Suen K, Leftheris K, Meyers CA, Barbacid M (1994) Vav cooperates with Ras to transform rodent fibroblasts but is not a Ras GDP/GTP exchange factor. Oncogene 9:2405-2413.

Chen Q, Kinch MS, Lin TH, Burridge K, Juliano RL (1994) Integrin-mediated cell adhesion activates mitogen-activated protein kinases. J Biol Chem 269:26602-26605.

Chen Q, Lin TH, Der CJ, Juliano RL (1996) Integrin-mediated activation of mitogen-activated protein (MAP) extracellular signal-related kinase kinase (MEK) and kinase is independent of Ras. J Biol Chem 271:18122-18127.

Clark EA, Hynes RO (1996) Ras activation is necessary for integrin-mediated activation of extracellular signal-regulated kinase 2 and cytosolic phospholypase A_2 but not for cytoskeletal organization. J Biol Chem 271:14814-14818.

Crespo P, Schubel KE, Ostrom AA, Gutkind JS, Bustelo XR (1997) Phosphotyrosine-dependent activation of Rac-1 GDP/GTP exchange by the vav proto-oncogene product. Nature 385:169-172.

Davis RJ (1994) MAPKs: new Jnk expands the group. Trends Biochem Sci 19:470-473.

Gulbin E, Coggeshall KM, Baier G, Katzav S, Burn P, Altman A (1993) Tyrosine kinase-stimulated guanine nucleotide exchange activity of vav in T cell activation. Science 260:822-825.

Kapron-Bras C, Fitz-Gibbon L, Jeevaratnam P, Wilkins J, Dedhar S (1993) Stimulation of tyrosine phosphorylation and accumulation of GTP-bound p21^{ras} upon antibody-mediated $\alpha2\beta1$ integrin activation in T-lymphoblastic cells. J Biol Chem 268:20701-20704.

Khosravi-Far R, Chrzanowska-Wodnicka M, Solki PA, Eva A, Burridge K, Der CJ (1994) Dbl and Vav mediate transformation via mitogen-activated protein kinase pathways that are distinct from those activated by oncogenic Ras. Mol Cell Biol 14:6848-6857.

Law DA, Nannizzi-Alaimo L, Phillips DR (1996) Outside-in integrin signal transduction. $\alpha_{IIb}\beta_3$-(GP IIb-IIIa) tyrosine phosphorylation induced by platelet aggregation. J Biol Chem 271:10811-10815.

Mainiero F, Murgia C, Wary KK, Pepe A, Blumemberg M, Westwick JK, Der CJ, Giancotti FG (1997) The coupling of $\alpha6\beta4$ integrin to Ras-MAPK pathways mediated by Shc controls keratinocyte proliferation. EMBO J 16:2365-2375.

Mainiero F, Pepe A, Wary KK, Spinardi L, Mohammadi M, Schlessinger J, Giancotti FG (1995) Signal transduction by the $\alpha6\beta4$ integrin: distinct $\beta4$ subunit sites mediate recruitment of Shc/GRB2 and association with the cytoskeleton of hemidesmosomes. EMBO J 14:4470-4481.

McCormick F (1993) How receptors turn Ras on. Nature 363:15-16.

McCormick F (1994) Raf: the holy grail of Ras biology? Trends Cell Biol 4:347-350.

Miyamoto S, Teramoto H, Coso OA, Gutkind JS, Burbelo PD, Akiyama SK, Yamada KM (1995) Integrin function: molecular hierarchies of cytoskeletal and signaling molecules. J Cell Biol 131:791-805.

Morino N, Mimura T, Hamasaki K, Tobe K, Ueki K, Kikuchi K, Takehara K, Kadowaki T, Yazaki Y, Nojima Y (1995) Matrix/integrin interaction activates the mitogen-activated protein kinase, p44^{erk-1} and p42^{erk-2}. J Biol Chem 270:269-273.

Renshaw MW, Toksoz D, Schwartz MA (1996) Involvement of the small GTPase Rho in integrin-mediated activation of mitogen-activated protein kinase. J Biol Chem 271:21691-21694.

Schlaepfer DD, Hanks SK, Hunter T, van der Geer P (1994) Integrin-mediated signal transduction linked to Ras pathway by GRB2 binding to focal adhesion kinase. Nature 372:786-791.

Schlaepfer DD, Hunter T (1996) Evidence for in vivo phosphorylation of the GRB2 SH2-domain binding site on focal adhesion kinase by Src-family protein-tyrosine kinases. Mol Cell Biol 16:5623-5633

Schlessinger J (1993) How receptor tyrosine kinases activate Ras. Trends Biochem Sci 18:273-275

Seger R, Krebs EG (1995) The MAPK signaling cascade. FASEB J 9:726-735

Su B, Karin M (1996) Mitogen-activated protein kinase cascades and regulation of gene expression. Curr Opin Immunol 8:402-411

Vojtek AB, Cooper JA (1995) Rho family members: activators of MAP kinase cascades. Cell 82:527-529

Wary KK, Mainiero F, Isakoff SJ, Marcantonio EE, Giancotti FG (1996) The adaptor protein Shc couples a class of integrins to the control of cell cycle progression. Cell 87:733-763.

Zheng L, Sjolander A, Eckerdal J, Andersson T (1996) Antibody-induced engagement of β2 integrins on adherent human neutrophils triggers activation of p21ras through tyrosine phosphorylation of the protooncogene product Vav. Proc Natl Acad Sci USA 93:8431-8436

Zhu X, Assoian RK (1995) Integrin-dependent activation of MAP kinase: a link to shape-dependent cell proliferation. Mol Biol Cell 6:273-282

Rho Family GTPases in Actin Cytoskeleton Assembly

Small GTPases belonging to the Rho family control cell growth and actin cytoskeleton organization. The effects on cell proliferation are likely to be due to activation of MAPK and are discussed in chapter 5. The ability of small GTPases to regulate actin cytoskeleton during adhesion and migration will be discussed here. Although little is known on the ability of integrins to trigger activation of small GTPases, their capacity to induce cytoskeleton organization during cell adhesion and locomotion, strongly suggests that small GTPases of the Rho family are part of the integrin signaling pathway.

Cell locomotion requires complex actin cytoskeleton reorganization and formation of membrane protrusions in the form of filopodia and lamellipodia. Filopodia are long, thin cylindrical structures generated by polymerization of actin in tightly-packed parallel bundles, while lamellipodia are large and flat membrane sheets formed by a orthogonally oriented web of actin filaments. Filopodia and lamellipodia are extended reversibly at the leading edge of locomoting cells and make contact with the surrounding extracellular matrix substratum. If, for any reason, contact with the substratum is not established, lamellipodia lift backward forming membrane ruffles. If contact is stabilized, small focal adhesions are formed progressively leading to the organization of actin stress fibers, whose contraction induce retraction of the rear edge of the cells (for review see Condeelis, 1993; Stossel, 1993; Small, 1994; Huttenlocher et al 1995; Mitchison and Cramer, 1996).

Rho family GTPases (Rho, Rac and Cdc42) can regulate focal adhesion formation and actin cytoskeleton organization. While Rho controls stress fibers assembly, Rac regulates formation of lamellipodia and ruffling, and Cdc42 filopodia formation (Hall, 1994; Takai et al 1995; Chant and Stowers, 1995; Machesky and Hall, 1996).

THE RHO FAMILY OF SMALL GTPASES

The Rho family belongs to the Ras superfamily of small GTPases and is composed by RhoA, B and C, Rac1 and 2, G25K/Cdc42, RhoG, RhoE and TC10 (Boguski and McCormick, 1993; Hall, 1994). One additional member, Rho L, has been identified in *Drosophila* (Murphy and Montell, 1996). These proteins share approximately 50% homology and are 30% identical to Ras in their amino acid sequence (Pai et al 1989; Self and Hall, 1995). RhoA, B and C share 85% homology to each other, with the divergence located mostly in the carboxyterminal

Signal Transduction by Integrins, by Paola Defilippi, Angela Gismondi, Angela Santoni and Guido Tarone. © 1997 Landes Bioscience.

part of the molecules. Rac1 and 2 are 92% identical to each other and G25K/Cdc42 are the most closely related isoforms, differing only in nine amino acid residues. The carboxyterminal protein region of the Rho family is characterized by the sequence CXXX, which directs a series of posttranslational modifications that include prenylation with a farnesyl (C15) or geranylgeranyl (C20) group to the cysteine residue. These modifications render the proteins hydrophobic and target them to the membrane (Boguski and McCormick, 1993; Philips and Pillinger, 1995).

The Rho family members bind GTP in the activated state (GTP-bound) and GDP in the inactive state (GDP-bound). The interconversion occurs through a cycle of guanine nucleotide exchange and GTP hydrolysis. Cycling between the active and inactive form is regulated by guanine nucleotide exchange factors (GEFs), GTPase-activating proteins (GAPs) and GDP-dissociation inhibitory factors (GDIs) (Table 6.1); GEFs activate Rho family members by accelerating their rate of GDP release, therefore facilitating their GTP binding. GDIs and GAPs are negative regulators of GTPases; the former inhibit GDP removal, while the latter stimulate the intrinsic GTP-hydrolyzing activity (Table 6.1) (Boguski and McCormick, 1993).

The biological role of the Rho family has been assessed by the use of mutated protein forms. The protein sequence is mutated in vitro to obtain constitutively active proteins or dominant negative forms. Mutation of Gly-14 to Val constitutively activates Rho (V14Rho) since it renders the molecule insensitive to GAP proteins and thus blocks Rho in the GTP-bound form. Dominant negative forms of Rho, Rac and Cdc42 have also been created by introducing mutations that inactivate the proteins. A mutated form of Rac, where Ser-17 has been converted to Asn (N17Rac), increases Rac affinity for GDP, thus behaving as a dominant negative form of Rac.

Rho family GTPases are also targets for covalent modification by many toxins produced by pathogenic bacteria. Toxins A and B from *Clostridium difficile* inactivate Rho and Cdc42 by glucosylation (Just et al 1995), C3 from *Clostridium botulinum* inactivates Rho by ADP-rybosylation (Rubin et al 1988), while Cytotoxic Necrotizing Factor 1 (CNF1) from *Escherichia coli* and *Bordetella bronchiseptica* Dermonecrotizing toxin (DNT) activate Rho by unknown mechanisms (Fiorentini et al 1995; Horiguchi et al 1995). C3 transferase is a 25 kDa exoenzyme produced by the bacterium *Clostridium botulinum*, which specifically ADP-ribosylates Rho on Asp-41. Since Asp-41 is located in the effector region of the molecule, it has been suggested that ADP-ribosylation disturbs the interaction of Rho with a putative effector, inactivating the protein (Rubin et al 1988; Braun et al 1989; Narumiya et al 1990). Due to its selectivity, C3 transferase has been widely used to test the biological activity of Rho.

EFFECTOR MOLECULES OF THE RHO FAMILY

Potential downstream targets of Rho family GTPases have been recently identified. They include Ser/Thr kinases, Tyr kinases, lipid kinases, phospholipases and lipoxygenases (Table 6.2).

Rho binds and activates in a GTP-dependent manner a family of related Ser/Thr kinases of 150-160 kDa, called Rho-kinases (indicated as Rok or Rock by different authors) (Leung et al 1995; Ishizaki et al 1996; Lamarche et al 1996; Leung et al 1996; Matsui et al 1996; Nakagawa et al 1996). Alignment of amino acids sequences shows that Rokα (Leung et al 1996) and p164 (Matsui et al 1996) represent the same molecule; Rokβ (Leung et al 1996) and p160Rock (Ishizaki et al 1996; Lamarche et al 1996) have identical amino acid sequences and are highly homologous to Rokα/p164. The proteins belonging to this family respond to activated Rho by translocating from the cytoplasm to the plasma membrane (Leung et al 1995; Matsui et al 1996). Although Rokβ/p160Rock has been identified as a Rho target, it may also be a target of Rac. Mutational analysis showed that amino acid residue Phe-37 of Rac is crucial

Table 6.1. GEFs, GAPs and GDIs for the Rho subfamily of small GTPases in mammalian cells

	Predicted molecular mass	Biochemical function	References
Abr	93 kDa	GEF for Cdc42>Rho>Rac GAP for Rac and Cdc42	Heisterkamp et al 1993; Tan et al 1993; Chuang et al 1995
Bcr	140 kDa	GEF for Cdc42>Rho>Rac GAP for Rac and Cdc42 (*)	Diekmann et al 1991; Ridley et al 1993; Chuang et al 1995
3 BP1	80 kda	GAP for Rac	Cicchetti et al 1995
α2-chimerin	45 kDa	Gap for Rac	Hall et al 1993
β-chimerin	34 kDa	GAP for Rac	Leung et al 1993
n-chimerin or α1-chimerin	34 kDa	GAP for Rac	Diekmann et al 1991; Amhed et al 1995
Dbl	115 kDa	GEF for Rho and Cdc42	Hart et al 1991; Hart et al 1994
Graf-2	65 kDa	GAP for Rho and Cdc42	Hildebrand et al 1996
Lbc	69 kDa	GEF for Rho	Zheng et al 1995
Ly-Gdi	27 kDa	GDI for Rho	Scherle et al 1993
Myr-5	225 kDa	GAP for Rho, Cdc42 and Rac	Reinhard et al 1995
Ost	100 kDa	GEF for Rho and Cdc42	Horii et al 1994
p190RhoGap	190 kDa	GAP for Cdc42>Rho and Rac (**)	Settleman et al 1992; Ridley et al 1993
p190-B	190 kDa	GAP for Rho, Rac and Cdc42	Burbelo et al 1995b
p122	122 kDa	GAP for Rho	Homma and Emori 1995
p50RhoGap	50 kDa	GAP for Cdc42>Rho and Rac	Lancaster et al 1994
RhoGdi	27 kDa	GDI for Rho, Rac and Cdc42	Ohga et al 1989; Hart et al 1992
RLIP76	76 kDa	GAP for Rac and Cdc42	Jullien-Flores et al 1995
SmgGds	60 kDa	GEF for Rho>Rac>Cdc42	Kaibuchi et al 1991; Chuang et al 1994
Tiam-1	177 kda	GEF for Rac and Cdc42	Michiels et al 1995
Trio	324 kDa	GEF for Rho and Rac	Debant et al 1996

> indicates that the former small GTPase is the preferred substrate.
(*) In vivo Bcr has a more specific GAP activity for Rac (Ridley et al 1993).
(**) In vivo p190RhoGAP is a GAP for Rho (Ridley et al 1993).

in Rokβ/p160Rock binding (Lamarche et al 1996). The Rho-kinases play an important role in Rho and Rac-dependent actin organization (see below).

Both Rac and Cdc42 bind the serine-theonine kinase p65Pak (Manser et al 1994), the prototype of a family of kinases (Pak family) which share the ability to bind to GTPases and to be activated following bind-ing (Bagrodia et al 1995; Manser et al 1995; Martin et al 1995). Activation of p65Pak requires both aminoterminal and carboxy-terminal regions of Rac (Diekmann et al 1995). Further analysis indicates that the amino acid residue Tyr-40 is crucial to p65Pak binding (Lamarche et al 1996). Analysis of the GTPase binding site of p65Pak and the related yeast protein Ste20

Table 6.2. Effectors of Rho family of small GTPases

Effector protein	Biochemical function	Rho	Rac	Cdc42	References
Pak family	Ser/Thr kinases	–	+	+	Manser et al 1994; Bagrodia et al 1995; Manser et al 1995; Martin et al 1995
PKN	Ser/Thr kinase	+	–	–	Amano et al 1996a; Watanabe et al 1996
Rokα/p164	Ser/Thr kinase	+	–	–	Leung et al 1996; Matsui et al 1996
Rokβ/p160Rock	Ser/Thr kinase	+	+	–	Leung et al 1996; Ishizaki et al 1996; Lamarche et al 1996
p125Fak (*)	Tyr kinase	+	–	–	Seckl et al 1995; Flinn and Ridley 1996
p120Ack	Tyr kinase	–	–	+	Manser et al 1993
PI-3 kinase	Lipid kinase	+	+	+	Kumagai et al 1993; Zhang et al 1993; Zhang et al 1995; Zheng et al 1994; Tolias et al 1995; Bokoch et al 1996
Ptdlns-5 kinase	Lipid kinase	+	–	–	Chong et al 1994
PLA2 (**)	Phospholipase	–	+	–	Peppenlebosch et al 1995
PLD (*)	Phospholipase	+	–	–	Bowman et al 1993; Malcolm et al 1994; Balboa and Thisel 1995; Schmidt et al 1996

(continued on next page)

Table 6.2. Effectors of Rho family of small GTPases (con't)

Effector protein	Biochemical function	Rho	Rac	Cdc42	References
p67Phox	NADPH oxidase enzyme complex	–	+	+	Abo et al 1991; Prigmore et al 1995
Citron	no enzymatic activity; unknown function	+	+	–	Madaule et al 1995
Por1	no enzymatic activity; unknown function	–	+	–	Van Aelst et al 1996
Rhophilin	no enzymatic activity; unknown function	+	–	–	Watanabe et al 1996
Rhotekin	no enzymatic activity; unknown function	+	–	–	Reid et al 1996
Wasp	Wiskott Aldrich syndrome protein; no enzymatic activity; unknown function	–	–	+	Aspentrom et al 1996; Kolluri et al 1996; Symons et al 1996
β-tubulin	cytoskeletal protein	–	+	–	Best et al 1996

+: ability, –: inability of Rho, Rac and Cdc42 to bind to the effector molecules.
(*) No direct interaction between the molecules has been proved so far.
(**) Activation of PLA2 by Rac is suggested by the fact that constitutively active Rac induces leukotriene synthesis (Peppenlebosch et al 1995).

has led to the identification of a minimal region of 16 amino acids required for Cdc42 and Rac-interactive binding (CRIB motif) (Burbelo et al 1995a). A data bank search of this sequence indicated that the CRIB motif is present in several eukaryotic proteins, including the uncharacterized mammalian protein MSE55 and a novel mammalian kinase MLK3 (Burbelo et al 1995a). The Pak family kinases are involved in Rac and Cdc42 ability to activate MAPK (see chapter 3).

Rho binds to the aminoterminal region of the 120 kDa protein kinase N (PKN) and activates PKN in a GTP-dependent manner (Amano et al 1996a; Watanabe et al 1996). PKN phosphorylation is blocked in presence of C3 transferase and occurs either by autophosphorylation or by the action of some Rho-associated kinases (Amano et al 1996a).

Indirect results indicate that Tyr kinases may be downstream effectors of Rho. A likely candidate is the cytoplasmic tyrosine kinase p125Fak (Ridley and Hall, 1994; Barry and Critchley, 1994). Treatment of cells with GTPγS, a nonhydrolyzable form of GTP which binds stably to G-proteins, induces p125Fak tyrosine phosphorylation, which is blocked by treating cells with C3 transferase (Seckl et al 1995). In addition, introduction of activated Rho (V14Rho) in mouse fibroblasts induces p125Fak tyrosine phosphorylation (Flinn and Ridley, 1996). Although no direct binding of Rho to p125Fak has been shown, p125FAK tyrosine phosphorylation may represent a molecular event downstream of Rho involved in stress fiber organization (see below).

Cdc42 binds in a GTP-dependent manner to a nonreceptor tyrosine kinase called p120Ack, which contains the CRIB motif. Binding of p120Ack to Cdc42 inhibits both the intrinsic and GAP-stimulated GTPase activity (Manser et al 1993), suggesting that p120Ack may counteract the effects of GAPs, thus sustaining the GTP-bound active form of Cdc42.

In addition to Tyr and Ser/Thr kinases, lipid kinases may be downstream effectors of small GTPases involved in the regulation of actin polymerization (see chapter 3 and below). PI-3 kinase may be activated by treating cells with GTPγS or recombinant Rho (Zhang et al 1993); the activation can be blocked by C3 transferase (Kumagai et al 1993; Zhang et al 1993; Zhang et al 1995).

Rac and Cdc42 also activate PI-3 kinase in a GTP-dependent manner and directly bind to the p85 subunit of PI-3 kinase, through its Rho-GAP homology domain (Zheng et al 1994; Tolias et al 1995; Bokoch et al 1996).

Rho can activate PtdIns-5 kinase, increasing the levels of PtdIns(4,5)P$_2$ (Chong et al 1994). Rho associates in vitro and coimmunoprecipitates in vivo with a 68 kDa PtdIns-5 kinase isoform in fibroblasts, indicating that the PtdIns-5 kinase may be a direct target of Rho (Ren et al 1996).

Rac is necessary for the activation of arachidonic acid metabolism, since expression of active V12Rac induces leukotriene synthesis (Peppenlebosch et al 1995). This finding suggests that phospholipase A2 (PLA2) and 5-lypooxygenase are Rac effectors.

Rho has also been reported to activate phospholipase D (PLD) in vitro and in vivo, although direct interaction between Rho and PLD has not been demonstrated (Bowman et al 1993; Malcolm et al 1994; Balboa et al 1995; Schmidt et al 1996).

In phagocytic cells, active Rac is required for the activation of the NADPH oxidase enzyme complex consisting of two membrane-bound cytochrome b558 subunits (p22 and gp91) and two proteins recruited from the cytosol, p47Phox and p67Phox (Abo et al 1991). p67Phox is the effector protein which directly binds Rac in two sites: the aminoterminal amino acids 30-40 and the carboxyterminal amino acids 143-175 (Diekmann et al 1995). p67Phox also binds to Cdc42 in neutrophils (Prigmore et al 1995).

Proteins devoid of enzymatic activity have also been found to bind Rho. These include Rhophilin (Watanabe et al 1996) and Rhotekin (Reid et al 1996), which share sequence homology with the Ser/Thr kinase

PKN discussed above. Although the physiological role of the interaction of Rho with PKN, Rhophilin and Rhotekin is still unclear, these molecules represent a family of Rho effectors proteins that bear a consensus Rho-binding sequence at the aminoterminal region.

Rho and Rac also bind to a protein of 183 kDa called "citron," identified by two-hybrid analysis, which contains leucine-zipper motifs, a proline rich region and a PH domain. On the basis of the sequence, citron, which is devoid of enzymatic activity, may be an adaptor protein (Madaule et al 1995).

Rac interacts in a GTP-dependent manner with a 34 kDa protein, called Por1 (partner of Rac) (Van Aelst et al 1996). Substitutions of Thr by Ala at position 35 in constitutively active V12Rac abolishes Rac binding to Por1, indicating that this amino acid residue is crucial for binding. Por1 is involved in Rac-mediated membrane ruffling.

Cdc42 binds to the Wiskott-Aldrich syndrome protein (Wasp), which contains a CRIB motif (Aspentrom et al 1996; Kolluri et al 1996; Symons et al 1996). Wasp was isolated by two-hybrid screening using Cdc42 as bait and by affinity chromatography as well as by coimmunoprecipitation from cell extracts (Aspentrom et al 1996; Kolluri et al 1996; Symons et al 1996).

Rac, but not Cdc42 or Rho, has also been shown to bind β-tubulin in the GTP-bound state (Best et al 1996).

RHO REGULATES STRESS FIBER FORMATION

Actin stress fibers are formed by contractile filament bundles of actin and myosin-II. They attach at one end into the plasma membrane at the focal adhesion and at the other end they insert in a second focal adhesion or in a meshwork of intermediate filaments that surrounds the nucleus. Focal contacts allow actin filaments to exert forces against the substratum. Stress fibers are quickly formed when tension is generated across the cells, i.e., by plating cells on rigid substrates coated with extracellular matrix proteins. They are rapidly disassembled when cells detach from the matrix or undergo mitosis. Within fibroblasts in tissues, stress fibers allow the cells to exert tension on the surrounding matrix.

The first indication that Rho might be involved in stress fibers organization comes from studies performed with the bacterial toxin C3 transferase able to specifically inhibit Rho. When C3 transferase is introduced into cells by microinjection or permeabilization, cells round up and actin stress fibers disassemble (Rubin et al 1988; Chardin et al 1989; Paterson et al 1990). Microinjection of the constitutively active V14Rho in serum-starved Swiss 3T3 fibroblasts induced stress fiber assembly (Fig. 6.1) (Ridley and Hall, 1992), thus demonstrating the regulatory role of Rho in this process. In other cells Rho controls additional actin-based structures and processes. In polarized epithelia, Rho is involved in the regulation of tight junctions and the organization of perijunctional actin (Nusrat et al 1995).

Rho-mediated stress fiber formation may also depend on Rac. As discussed above Rac generates leukotrienes, which are necessary and sufficient for stress fiber formation in mouse Swiss 3T3 fibroblasts (Peppelenbosch et al 1995). Leukotriene-induced stress fiber assembly is blocked by C3 transferase, indicating that this process requires Rho. Thus it is possible that Rac is upstream to Rho in one of the pathways that trigger stress fiber formation (see also below).

Amino acid substitutions at positions 34 and 36 of Rho block induction of stress fibers, indicating that the aminoterminal region of Rho is an effector domain required for stress fiber formation (Self et al 1993).

A possible mechanism by which Rho can affect stress fiber organization in nonmuscle cells is stimulation of cell contractility mediated by myosin light chain phosphorylation. In vitro and in vivo analysis show that Rho can bind to Ser/Thr Rokα/p164 kinase (Matsui et al 1996), which phosphorylates

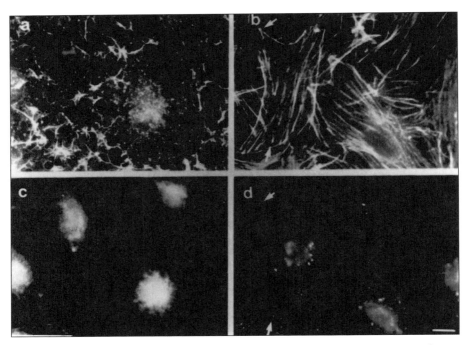

Fig. 6.1. Actin filaments and vinculin distribution in constitutively active V14Rho-injected Swiss 3T3 cells. Cells were serum starved to induce disassembly of actin stress fibers and focal adhesions. Cells were fixed before (a and c) or 30 min after injection with V14Rho at 300 μg/ml (b and d). Actin filaments were visualized with FITC-conjugated phallacidin (a and b) and in the same cells vinculin was localized by immunofluorescence with a mouse anti-vinculin antibody (c and d). Vinculin localizes with the ends of stress fibers, e.g., pairs of arrows in (b) and (d). Bar represents 10 μm. With permission from Ridley AJ, Hall A. Cell 1992; 70:389-399. Copyright Cell Press.

myosin light chain, facilitating the actin activation of myosin ATPase (Amano et al 1996b). At the same time Rokα/p164 kinase phosphorylates and inactivates the myosin light chain phosphatase (Kimura et al 1996). Thus, Rho may trigger cell contractility by controlling the level of phosphorylation of myosin light chain both directly or by inhibiting myosin light chain phosphatase. Consistent with the pathway described above, inhibition of myosin light chain phosphorylation by different kinase inhibitors prevents stress fiber formation in cells microinjected with activated Rho (Chrzanowska and Burridge, 1996).

A direct role for the Rokα/p164 kinase in the reorganization of the cytoskeleton has also been recently described (Leung et al 1996). Microinjection of full length Rokα/p164 kinase cDNA promotes formation of stress fibers and focal adhesions in HeLa cells, which also occurs in the presence of C3 transferase, indicating that this protein acts downstream of Rho (Leung et al 1996). Mutational analysis shows that both the aminoterminal region and the kinase activity of Rokα/p164 kinase are necessary to drive stress fiber formation (Fig. 6.2). The aminoterminal truncated form, rather than promoting formation of stress fibers, induces loss of these structures, indicating that different domains of the molecule can exert regulatory functions on stress fiber formation.

Another mechanism by which Rho may trigger stress fiber formation is induction of

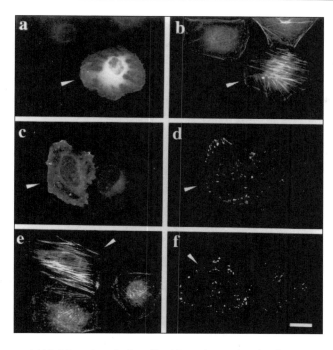

Fig. 6.2. Overexpression of Roka/p164 promotes formation of stress fibers and focal adhesions in HeLa cells. Cells were serum starved for 24 hours to induce disassembly of actin stress fibers and focal adhesions and microinjected with DNA constructs encoding full length Roka/p164 (a to d) or carboxyterminally truncated Roka/p164 (c and f) and examined 2 h later. All constructs used in this study contained N-terminally hemoagglutinin (HA)-tagged sequence. Cells were double stained with anti-HA monoclonal antibody 12CA5 (a) to detect protein expression and phalloidin (b) to detect actin filaments or with rhodamine-conjugated anti-HA (c) and anti vinculin (d) to detect focal adhesions. In panels e and f cells were stained with phalloidin and anti-vinculin antibody, respectively. Arrowheads point to the injected cells. Bars represents 10. With permission from Leung T et al Mol Cell Biol 1996; 16:5313-5327.

p125Fak tyrosine phosphorylation (see also chapter 2). Activated Rho can induce p125Fak tyrosine phosphorylation (Flinn and Ridley, 1996), suggesting that this event may be involved in Rho-induced stress fibers assembly.

Rho-mediated activation of PtdIns-5 kinase and production of $PtdIns(4,5)P_2$ has been proposed as an additional mechanism by which Rho can regulate assembly of focal adhesions and stress fibers. It has been shown that the carboxyterminal domain of vinculin contains an F-actin binding site, which is normally masked by a head-tail interaction. When head-tail connections are disrupted, the vinculin molecules are open and display higher binding affinity for talin and actin (Johnson and Craig, 1995). Recent results demonstrate that $PtdIns(4,5)P_2$ is able to dissociate vinculin head-tail interaction, activating the molecule to bind talin

and actin (Gilmore and Burridge, 1996; Weekes et al 1996). Cells microinjected with antibodies against $PtdIns(4,5)P_2$ do not form focal adhesions and stress fibers, indicating that $PtdIns(4,5)P_2$ is required in the assembly of focal adhesions (Fig. 6.3) (Gilmore and Burridge, 1996).

A summary of the pathways involved in Rho-mediated focal adhesion and stress fiber formation is shown in Figure 6.4.

RAC-MEDIATED LAMELLIPODIA FORMATION

Formation of lamellipodia and filopodia requires polymerization of actin filaments. Actin filaments are polar structures with two structurally different ends, a "pointed" and a "barbed" end. Addition of monomeric actin to the growing filaments occurs only on the barbed end, which is usually capped by proteins that do not allow addition of

Fig. 6.3. Sequestering PtdIns(4,5)P$_2$ prevents formation of stress fibers and focal adhesions in Balb/c 3T3 cells. Cells were serum starved to induce disassembly of stress fibers and focal adhesions. Starved cells were microinjected with anti-PtdIns(4,5)P$_2$ monoclonal antibody and then stimulated for 20 min with 0.5% fetal bovine serum to induce stress fibers and focal adhesion assembly. Starved cells stained for actin filaments (a) and vinculin (b). Cells stimulated with fetal bovine serum following antibody injection (c-j). Antibodies were coinjected with coumarin-conjugated bovine serum albumin to identify injected cells (d, f, h and j). Cells injected with control antibody, stained for actin filament (c). Cells injected with anti-PtdIns(4,5)P$_2$ monoclonal antibody and stained for actin filament (e), talin (g) or vinculin (i). Reprinted with permission from Gilmore AP, Burridge K Nature 1996; 381:531-535. Copyright 1996 Macmillan Magazines Limited.

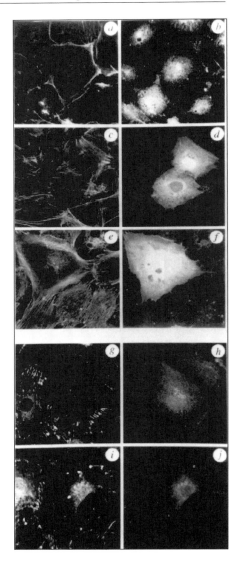

monomeric actin. Uncapping of barbed ends occurs in response to stimuli that elicit actin polymerization (Hartwig, 1992; Condeelis, 1993; Sun et al 1995). Subsequently actin filaments can assemble either in a tridimensional network that support lamellipodia or in tightly packed parallel bundles forming filopodia.

The first insights on Rac control of actin polymerization were provided in platelets, which, in response to injury, are activated and undergo shape transformations. Activated platelets expand into a sphere and develop surface protrusions, flat lamellae and finger-like filopodia, which allow adhesion to injured surfaces and formation of tridimensional clots. Platelet configuration is regulated by remodeling of actin cytoskeleton (Hartwig, 1992; Fox et al 1993). Using permeabilized platelets Hartwig et al (1995) showed that constitutively active Rac, but not Rho, activates uncapping of actin barbed ends through phosphoinositide synthesis. Since Cdc42 and Rho have also been shown to trigger production of polyphosphoinositides (Zhang et al 1993; Chong et al 1994; Zheng et al 1994), these GTPases may also use similar mechanisms to control

filament uncapping in other cellular systems.

Microinjection of active V12Rac in serum-starved Swiss 3T3 fibroblasts stimulates actin filament accumulation at the plasma membrane inducing lamellipodia and membrane ruffling (Fig. 6.5), (Ridley et al 1992). Following initial accumulation of actin in membrane ruffles, microinjected V12Rac also leads to stress fiber formation, which was shown to depend on Rac-induced

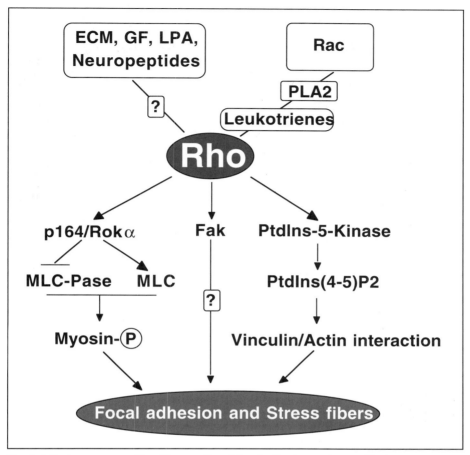

Fig. 6.4. Pathways controlling Rho-mediated focal adhesion and stress fibers organization. Rho can be activated by extracellular stimuli that include matrix proteins (ECM), growth factors (GF), lysophosphatidic acid (LPA) or neuropeptides or by Rac via phospholipase A2 and leukotrienes production. Rho can than activate multiple ways leading to organization of actin fibers. These include: 10 the activation of the Rokα/p164 kinase that can activate myosin by phosphorylating the myosin light chain (MLC) and by inhibiting the MLC phosphatase (MLC-Pase); 2) activation of p125Fak that controls actin fibers organization by still unknown mechanisms (see chapter 2); and 3) activation of the PtdIns-5 kinase responsible for the synthesis of the PtdIns(4,5)P$_2$ that in turn can activate vinculin to interact with actin and thus, promote focal adhesion and stress fiber organization.

leukotrienes synthesis, which in turn activates Rho (Ridley et al 1992; Peppelembosch et al 1995). These data are consistent with a model in which activated Rac induces membrane ruffling and also leads to Rho stimulation and subsequent stress fiber formation (see Fig. 6.4).

The ability of Rac to trigger actin polymerization in lamellipodia depends on an aminoterminal effector site located at amino acids 30-40, while a second effector site has been located at amino acids 143-175 (Diekmann et al 1995). These effector sites are also essential for activation of the Ser/Thr Pak65 kinase. Experiments performed with constitutively active Rac containing selected amino acids substitutions in the aminoterminal effector region,

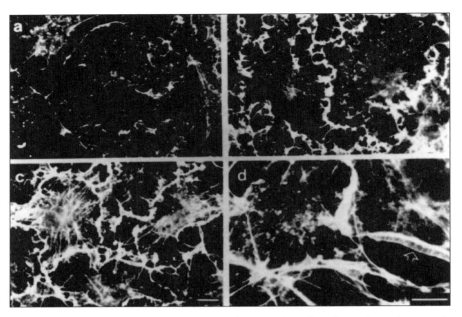

Fig. 6.5. Rac induces actin filament accumulation in lamellipodia and membrane ruffling in Swiss 3T3 cells. Cells were serum starved to induce disassembly of actin stress
fibers and focal adhesions. Purified recombinant constitutively active V12Rac at 300
µg/ml was microinjected and cells were fixed after 15 (b) or 30 min (c and d). Cells
were stained with TRITC-labeled phalloidin to show actin filaments. The appearance
of actin in lamellipodia and membrane ruffling is clearly shown at higher resolution (d,
arrow). Control uninjected cells are shown in (a). Bars (c and d) represent 10 µM. With
permission from Ridley AJ et al Cell 1992; 70:401-410. Copyright Cell Press.

allowed classification of the target proteins
responsible for Rac biological activities.
Phe37Ala Rac mutant is unable to induce
actin polymerization in lamellipodia and
focal adhesion formation. Since the
Phe37Ala mutant does not bind to
Rokβ/p160Rock (Lamarche et al 1996) and
to Por1 (Joneson et al 1996), these proteins
may provide an essential signal required for
Rac-induced actin assembly. A truncated
form of Por1, which retains the ability to
bind Rac, inhibits induction of Rac-dependent lamellipodia and membrane ruffling,
further supporting a role of Por1 in Rac-induced actin reorganization (Van Aelst et al
1996). On the other hand, mutation of
Tyr40Cys residue of Rac does not affect
lamellipodia formation but blocks activa-

tion of p65Pak kinase, thus showing that this
Ser/Thr kinase is not involved in Rac-induced actin assembly.

In vivo, Drac1, a *Drosophila* homolog of
Rac, has been shown to be involved in morphogenesis. Overexpression of constitutively active Drac1 in the *Drosophila* nervous system causes axon outgrowth defects
in peripheral neurons without affecting den-
drites, while overexpression in muscle precursors blocks myoblast fusion (Luo et al
1994). Moreover, *Drosophila* Rac1 controls
the assembly of actin at adherens junction
of the wing disc epithelium (Eaton et al
1995). In transgenic mice expression of constitutively active Rac in Purkinje cells has a
differential effect on axons, dendritic trunks
and spines (Luo et al 1996).

CDC42 CONTROLS THE FORMATION OF FILOPODIA

Microinjection of Cdc42 in subconfluent serum-starved Swiss 3T3 fibroblasts leads to reorganization of actin in long peripheral extensions, indicating that Cdc42 regulates organization of actin in filopodia (Fig. 6.6) (Kozma et al 1995; Nobes and Hall, 1995). Similar results were also obtained by expressing constitutively activated Cdc42 in human HeLa cells (Dutartre et al 1996). Prolonged incubation of cells expressing constitutively active Cdc42 leads to lamellipodia and stress fiber formation (Nobes and Hall, 1995). These effects are inhibited by dominant negative Rac and C3 transferase, suggesting that Cdc42 can sequential activate Rac and Rho (Ridley et al 1992). These findings implicate that, at least in Swiss 3T3 fibroblasts, a sequential activation cascade exists between these three small GTPases. Interestingly in the study of Kozma et al (1995), microinjection of wild type Cdc42 induces filopodia and lamellipodia but causes disassembly of stress fibers. Thus, Cdc42 can activate Rac while the ability to sequentially activate Rho may depend on the experimental conditions used, as was observed in cells microinjected with constitutively active (Nobes and Hall, 1995) but not with the wild type form of Cdc42 (Kozma et al 1995).

Downstream effectors responsible for filopodia organization are not known. Mutation in Phe-37 blocks the ability of Cdc42 to induce lamellipodia via Rac activation, indicating that this Cdc42 site binds an unknown effector molecule important in the activation of the GTPase cascade (Lamarche et al 1996).

Cdc42-expressing cells block cytokinesis and become large and multinucleate (Dutartre et al 1996). Giant and multinucleate cells have also been described upon expression of Dbl or Tiam-1, two known exchange factors for Cdc42 (Hart et al 1991; Ron et al 1991; Habets et al 1994; Michiels et al 1995). These data indicate that activation of Cdc42, either by expressing constitutively active forms of Cdc42 or overexpressing GDP/GTP exchange factors, interferes with the cell division process.

Increased expression and plasma membrane association of Cdc42 occurs during monocyte-macrophages differentiation by phorbol esters, a process which leads to cell adhesion and spreading. Cdc42 is not regulated during differentiation with agonists that do not enhance cell adhesion (Aepfelbacher et al 1994) suggesting a correlation between levels of Cdc42 in the plasma membrane and the ability of cells to spread.

Cdc42 has also been shown to regulate polarization in helper T lymphocytes, localizing actin at the site of cell-cell contact during antigen presentation (Stowers et al 1995).

Recently it has been shown that Cdc42 interacts with the Wiskott-Aldrich syndrome protein (Wasp), which contains a CRIB motif (Aspentrom et al 1996; Kolluri et al 1996; Symons et al 1996). The Wiskott-Aldrich syndrome is an X-linked immunodeficiency disorder which affects T lymphocytes, neutrophils and platelets. Affected males present abnormal small platelets and thrombocytopenia; their T lymphocytes show actin cytoskeletal abnormalities and their neutrophils are defective in chemotactic migration, suggesting that defects in actin cytoskeleton may lie at the basis of this syndrome (Ochs et al 1980; Kenney et al 1986; Molina et al 1992; Derry et al 1994). These data suggest that Wasp is an important downstream effector implicated in Cdc42-dependent actin polymerization in hemopoietic cells.

GEFS AND GAPS FOR SMALL GTPASES ALSO REGULATE ACTIN CYTOSKELETON ASSEMBLY

As mentioned above, GTP binding to small GTPases is regulated by guanine exchange factors GEFs, while GTPase activity is stimulated by GAPs (Fig. 6.7 and Table 6.1). It is conceivable that overexpression of GEFs mimics activation of small GTPases, while overexpression of GAPs can inhibit their activity.

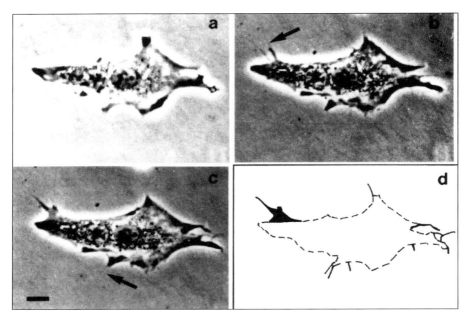

Fig. 6.6. Cdc42 promotes the formation of filopodia in Swiss 3T3 fibroblasts. Cells were serum starved to induce disassembly of actin stress fibers and focal adhesions. Wild type Cdc42 (500 μg/ml) was microinjected into subconfluent fibroblasts, which were photographed under phase-contrast microscopy at 0 (a), 5 (b) and 15 (c) min after microinjection. Arrows (b and c) indicate filopodia. (d) Diagrammatic representation showing cell shape at 0 min (dashed lines), new filopodia present at 15 min (black lines extending out from cell) and areas of cell extension as lamellipodia (filled-in areas). With permission from Kozma R et al Mol Cell Biol 1995; 15:1942-1952.

Tiam-1 is a GEF for Rac and Cdc42. Expression of Tiam-1 in NIH3T3 cells leads to actin organization in lamellipodia and membrane ruffling, which is inhibited by coexpression of dominant negative N17Rac (Michiels et al 1995). These results indicate that Tiam-1-induced actin polymerization is mediated by activation of Rac. Tiam-1 was initially identified as an oncogene able to induce invasive phenotype in T lymphoma cells (Habets et al 1994). Overexpression of constitutively active Rac produces a similar oncogenic phenotype, suggesting that the oncogenic effects of Tiam-1 are a consequence of Rac activation (Michiels et al 1995).

Cells overexpressing Dbl, a GEF for Cdc42 and Rho (Hart et al 1991; Hart et al 1994), form stress fibers on fibronectin, and organize filopodia and lamellipodia when plated on collagens or gelatin, suggesting

that Dbl may differentially affect cytoskeletal organization in response to different matrix proteins (Defilippi et al 1997).

p190Rho-Gap, α-chimerin and 3 BP-1 are GAPs for small GTPases that profoundly affect actin cytoskeleton organization. p190Rho-Gap (Settleman et al 1992) binds to a truncated form of Ras GAP (GAP-N). The complex GAP-N/p190Rho-Gap has a GAP activity on Rho. GAP-N expression in cells correlates with changes in cell cytoskeleton, resulting in loss of stress fibers, reduction of focal contacts and impaired ability of cells to adhere to fibronectin (McGlade et al 1993). Since these effects are similar to those observed in cells after downregulation of Rho, it is possible that the GAP-N-p190Rho-Gap complex has enhanced Rho GAP activity, thereby down regulating Rho. Cells expressing α-chimerin (also indicated as n-chimerin), a GAP for

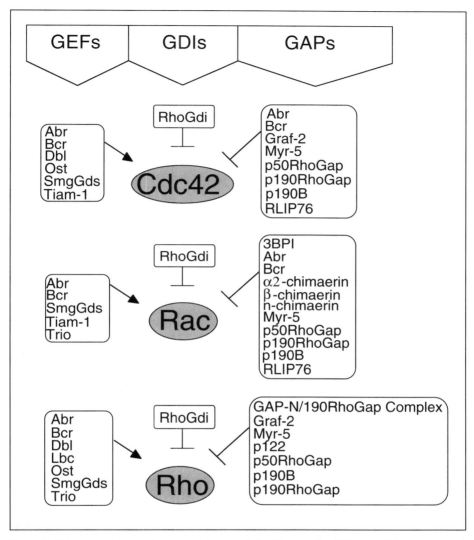

Fig. 6.7. Diagram of the various modulators of the Rho family GTPases: Cdc42, Rac and Rho. GEFs: guanine nucleotide exchange factors; GDIs: GDP dissociation inhibitors; GAPs: GTPase activating proteins.

Rac (Diekmann et al 1991; Ahmed et al 1995), do not assemble stress fibers and focal adhesion on fibronectin (Herrera and Shivers, 1994), suggesting that expression of this GAP may negatively regulate Rac activity, leading to altered adhesion-induced cytoskeletal assembly. 3 BP-1, a GAP for Rac, also interferes with small GTPases activity. In Swiss 3T3 cells microinjection of 3 BP1

GAP domain blocks Rac-induced lamellipodia formation and membrane ruffling (Cicchetti et al 1995).

Another possible connection between GAPs for small GTPases and cytoskeleton has been found in a novel myosin form, Myr 5 (*my*osin from *r*at). Myr 5 is able to bind actin in an ATP-regulated manner and has GAP activity for Rho, Rac and Cdc42 in vitro (Reinhard et al 1995), thus representing a possible regulatory molecule directly linking small GTPases and actin cytoskeleton.

REGULATION OF INTEGRIN-MEDIATED ADHESION

Small GTPases can also affect integrin-dependent cell adhesion and focal contact organization.

The first indication that Rho may trigger integrin-mediated adhesion comes from experiments in platelets, where C3 transferase blocks αIIb/β3 integrin-mediated platelet aggregation (Morii et al 1992). Similar results were obtained in lymphocytes, where PMA-induced β2 integrin-dependent aggregation was blocked by treating cells with C3 transferase (Tominaga et al 1993). Inactivation of Rho by C3 transferase also blocks FMLP or IL-8-induced binding of lymphocyte α4β1 integrin to VCAM-1 and of neutrophils β2 integrin to fibrinogen (Laudanna et al 1996). In these cells FMLP and IL-8 increase the amount of GTPγS bound to Rho, indicating that FMLP- and IL-8-induced α4β1 and β2 integrin-mediated adhesion requires activation of Rho.

In contrast to the data reported above indicating that Rho can activate integrin-mediated adhesion, inhibition of Rho activity by C3 transferase increases α5β1-dependent adhesion to fibronectin in U937 monocytic cells (Aepfelbacher, 1995), suggesting that Rho differently regulates integrin-mediated adhesion in different cell types.

The mechanism by which Rho can regulate integrin-mediated adhesion in platelets or leukocytes is not known. It can be postulated that Rho is involved in regulation of integrin affinity and/or avidity by controlling their clustering in the plasma membrane or their interaction with cytoskeletal elements. Indications for such mechanisms come from a study with R-Ras, a small GTPase that, as Rho, Rac and Cdc42, belongs to the Ras superfamily. Expression of a constitutively active R-ras converted two cell lines that grow in suspension into highly

adherent cells. Integrin ligands bind to these cells with higher affinities, indicating that R-Ras controls integrin-ligand binding; this finding suggests the possibility that the effect of Rho on adhesion may be mediated by similar mechanisms (Zhang et al 1996).

Small GTPases may also be important in controlling integrin clustering in focal contacts. In serum-starved Swiss 3T3 fibroblasts, interaction of integrins with fibronectin-coated dishes is not sufficient to induce focal contacts and their formation requires active Rho (Hotchin and Hall, 1995). The proposed model indicates that formation of focal contacts requires two signals: one provided by integrin-matrix interaction and the second by functionally active Rho, which may be required to cluster integrins on the cell surface.

In Swiss 3T3 fibroblasts, whose actin cytoskeleton and focal adhesions have been disassembled by serum starvation, microinjection of V14Rho or V12Rac induces assembly of vinculin-containing focal complexes (see Fig. 6.1) (Ridley and Hall, 1992; Nobes and Hall, 1995). V12Rac microinjection leads to focal adhesion assembly also in presence of C3 transferase, indicating that focal adhesions formed by Rac are independent from Rho. In the same set of experiments, microinjection of V12Cdc42 in the presence of C3 transferase and of N17Rac induces formation of focal contacts along and at the tip of filopodia (Nobes and Hall, 1995), thus, indicating that the complexes formed by Cdc42 are formed independently from those induced by Rac or Rho. These data suggest that Cdc42, Rac and Rho are each able to lead to organization of focal complexes through distinct signaling pathways. It is likely that the focal complexes formed by Cdc42 and Rac represent primordial adhesive structures localized at the tips of filopodia and lamellae, which may play a role for the assembly of stress fibers.

A possible mechanism by which Rho can lead to formation of focal complexes is via stimulation of PtdIns(4,5)P_2 synthesis (see above and Fig. 6.4). Support for the involvement of Rho in the integrin-mediated rise in PtdIns(4,5)P$_2$ levels comes from experiments in a gastric tumor cell line, where β1 integrin-mediated adhesion to collagen leads to production of PtdIns(4,5)P$_2$ blocked by treatment with C3 transferase (Udagawa and McIntyre, 1996).

Proteins with GAP activity are also present in the adhesive structures and can contribute to their turnover. Following adhesion to fibronectin-coated beads, p190 B, a new member of p190Rho-Gap family, clusters to adhesion sites, suggesting that this regulatory molecule may be involved in mediating integrin-associated changes in the actin cytoskeleton (Burbelo et al 1995b). Moreover, Graf-2, a novel GTPase-activating protein whose activity is directed to Cdc42 and Rho, associates in vitro with p125Fak in an SH3 domain dependent-manner (Hildebrand et al 1996) (see also chapter 2). Overexpression of Graf-2 in embryo fibroblasts reduces the number of actin stress fibers, consistent with a Rho GAP activity (Parsons, 1996). Graf-2 binding to p125Fak suggests a crosstalk between tyrosine kinases and Rho family GTPases and may represent a regulatory mechanism of focal adhesion disassembly.

REFERENCES

Abo A, Pick E, Hall A, Totty N, Teahan CG, Segal AW (1991) Activation of the NADPH oxidase involves the small GTP-binding protein p21rac1. Nature 353:668-670.

Aepfaelbacher M (1995) ADP-ribosylation of Rho enhances adhesion of U937 cells to fibronectin via the alpha5beta1 integrin receptor. FEBS Lett 363:78-80.

Aepfelbacher M, Vauti F, Weber PC, Glomset JA (1994) Spreading of differentiating human monocytes is associated with a major increase in membrane-bound CDC42. Proc Natl Acad Sci USA 91:4263-4267.

Ahmed S, Kozma R, Hall C, Lim L (1995) GTPase-activating protein activity of n(alpha 1)-chimerin and effect of lipids. Methods Enzymol 256:114-25.

Amano M, Mukai H, Ono Y, Chihara K, Matsui T, Hamajima Y, Okawa K,

Iwamatsu A, Kaibuchi K (1996a) Identification of a putative target for Rho as the serine-threonine kinase protein kinase N. Science 271:648-650.

Amano M, Ito M, Kimura K, Fukata Y, Chihara K, Nakano T, Matsuura Y, Kaibuchi K (1996b) Phosphorylation and activation of myosin by Rho-associated kinase (Rho-kinase). J Biol Chem 271:20246-20249.

Aspenstrom P, Lindberg U, Hall A (1996) Two GTPases, Cdc42 and Rac, bind directly to a protein implicated in the immunodeficiency disorder Wiskott-Aldrich syndrome. Curr Biol 6:70-75.

Bagrodia S, Taylor SJ, Creasy CL, Chernoff J, Cerione RA (1995) Identification of a mouse p21Cdc42/Rac activated kinase. J Biol Chem 270:22731-22737.

Balboa MA, Insel PA (1995) Nuclear phospholipase D in Madin-Darby canine kidney cells. Guanosine 5'-O-(thiotriphosphate)-stimulated activation is mediated by rhoA and is downstream of protein kinase C. J Biol Chem 270:29843-29847.

Barry ST, Critchley DR (1994) The Rho-A dependent assembly of focal adhesions in Swiss 3T3 cells is associated with increased tyrosine phosphorylation and the recruitment of p125FAK and protein kinase C-δ to focal adhesions. J Cell Sci 107:2033-2045.

Best A, Ahmed S, Kozma R, Lim L (1996) The Ras-related GTPase Rac1 binds tubulin. J Biol Chem 271:3756-3762.

Boguski MS, McCormick F (1993) Proteins regulating Ras and its relatives. Nature 366:643-654.

Bokoch GM, Vlahos CJ, Wang Y, Knaus UG, Traynor-Kaplan AE (1996) Rac GTPase interacts specifically with phosphatidylinositol 3-kinase. Biochem J 315 (Pt 3):775-779.

Bohmer RM, Scharf E, Assoian RK (1996) Cytoskeletal integrity is required throughout the mitogen stimulation phase of the cell cycle and mediates the anchorage-dependent expression of cyclin D1. Mol Biol Cell 7:101-111.

Bowman EP, Uhlinger DJ, Lambeth JD (1993) Neutrophil phospholipase D is activated by a membrane-associated Rho family small molecular weight GTP-binding protein. J Biol Chem 268:21509-21512.

Braun U, Habermann B, Just I, Aktories K, Vandekerckhove J (1989) Purification of the 22 kDa protein substrate of botulinum ADP-ribosyltransferase C3 from porcine brain cytosol and its characterization as a GTP-binding protein highly homologous to the rho gene product. FEBS Lett 243:70-76.

Burbelo PD, Drechsel D, Hall A (1995a) A conserved binding motif defines numerous candidate target proteins for both Cdc42 and Rac GTPases. J Biol Chem 270:29071-29074.

Burbelo PD, Miyamoto S, Utani A, Brill S, Yamada KM, Hall A, Yamada Y (1995b) p190-B, a new member of the Rho GAP family, and Rho are induced to cluster after integrin crosslinking. J Biol Chem 270:30919-30926.

Chant J, Stowers L (1995) GTPase cascades choreographing cellular behaviour: movement, morphogenesis, and more. Cell 81:1-4.

Chardin P, Boquet P, Madaule P, Popoff MR, Rubin EJ, Gill DM (1989) The mammalian G protein Rho is ADP-ribosylated by Clostridium botulinum exoenzyme C3 and affects actin microfilaments in Vero cells. EMBO J 8:1087-1092.

Chen Q, Kinch MS, Lin TS, Burridge K, Juliano RL (1994) Integrin-mediated cell adhesion activates mitogen-activated protein kinases. J Biol Chem 269:26602-26605.

Chong LD, Traynor-Kaplan A, Bokoch GM, Schwartz MA (1994) The small GTP-binding protein rho regulates a phosphatidylinositol 4-phosphate 5-kinase in mammalian cells. Cell 79:507-513.

Chrzanowska-Wodnicka M, Burridge K (1996) Rho-stimulated contractility drives the formation of stress fibers and focal adhesions. J. Cell Biol 133:1403-1415.

Chuang TH, Xu X, Kaartinen V, Heisterkamp N, Groffen J, Bokoch GM (1995) Abr and Bcr are multifunctional regulators of the Rho GTP-binding protein family. Proc Natl Acad Sci USA 92:10282-10286.

Chuang TH, Xu X, Quilliam LA, Bokoch GM (1994) SmgGDS stabilizes nucleotide-bound and -free forms of the Rac1 GTP-binding protein and stimulates GTP/GDP

exchange through a substituted enzyme mechanism. Biochem J 303:761-767.

Cicchetti P, Ridley AJ, Zheng Y, Cerione RA, Baltimore D (1995) 3 BP-1, an SH3 domain binding protein, has GAP activity for Rac and inhibits growth factor-induced membrane ruffling in fibroblasts. EMBO J 14:3127-3315.

Condeelis J (1993) Life at the leading edge. Annu Rev Cell Biol 9:411-444.

Debant A, Serra-Pages C, Seipel K, O'Brien S, Tang M, Park SH (1996) The multidomain protein Trio binds the LAR transmembrane tyrosine phosphatase, contains a protein kinase domain, and has separate rac-specific and rho-specific guanine nucleotide exchange factor domains. Proc Natl Acad Sci USA 93:5466-5471.

Defilippi P, Olivo C, Tarone G, Mancini P, Torrisi MR, Eva A (1997) Actin cytoskeleton polymerization in Dbl-transformed NIH/3T3 fibroblasts is dependent on cell adhesion to specific extracellular matrix proteins. Oncogene 14:1933-1943.

Derry JM, Ochs HD, Francke U (1994) Isolation of a novel gene mutated in Wiskott-Aldrich syndrome. Cell 78:835-844.

Diekmann D, Brill S, Garrett MD, Totty N, Hsuan J, Monfries C, Hall C, Lim L, Hall A (1991) Bcr encodes a GTPase-activating protein for p21rac. Nature 351:400-402.

Diekmann D, Nobes CD, Burbelo PD, Abo A, Hall A (1995) Rac GTPase interacts with GAPs and target proteins through multiple effector sites. EMBO J 14:5297-5305.

Dutartre H, Davoust J, Gorvel JP, Chavrier P (1996) Cytokinesis arrest and redistribution of actin-cytoskeleton regulatory components in cells expressing the Rho GTPase CDC42Hs. J Cell Sci 109:367-377.

Eaton S, Auvinen P, Luo L, Jan YN, Simons K (1995) Cdc42 and Rac control different actin-dependent processes in the Drosophila wing epithelium. J Cell Biol 131:151-164.

Fiorentini C, Donelli G, Matarrase P, Fabbri A, Paradisi S, Boquet P (1995) *Escherichia Coli* cytotoxic necrotizing factor 1: evidence for induction of actin assembly by constitutive activation of the p21 Rho GTPase. Infect Immun 63:3936-3944.

Flinn H M, Ridley A J (1996) Rho stimulates tyrosine phosphorylation of focal adhesion kinase, p130 and paxillin. J Cell Science, 109:1133-1141.

Fox JEB (1993) The platelet cytoskeleton. Thromb Haemost 70:884-893.

Gilmore AP, Burridge K (1996) Regulation of vinculin binding to talin and actin by phosphatidyl-inositol-4-5-bisphosphate. Nature 381:531-535.

Habets GG, Scholtes EH, Zuydgeest D, van der Kammen RA, Stam JC, Berns A, Collard JG (1994) Identification of an invasion-inducing gene, Tiam-1, that encodes a protein with homology to GDP-GTP exchangers for Rho-like proteins. Cell 77:537-549.

Hall C, Sin WC, Teo M, Michael GJ, Smith P, Dong JM, Lim HH, Manser E, Spurr NK, Jones TA et al (1993) Alpha 2-chimerin, an SH2-containing GTPase-activating protein for the ras-related protein p21rac derived by alternate splicing of the human n-chimerin gene, is selectively expressed in brain regions and testis. Mol Cell Biol 13:4986-4998.

Hall A (1994) Small GTP-binding proteins and the regulation of the actin cytoskeleton. Annu Rev Cell Biol 10:31-54.

Hart MJ, Eva A, Evans T, Aaronson SA, Cerione RA (1991) Catalysis of guanine nucleotide exchange on the CDC42Hs protein by the dbl oncogene product. Nature 354:311314.

Hart MJ, Maru Y, Leonard D, Vitte ON, Evans T, Cerione RA (1992) A GDP dissociation inhibitor that serves as a GTPase inhibitor for the Ras-like protein Cdc42Hs. Science 258:812-815.

Hart MJ, Eva A, Zangrilli D, Aaronson SA, Evans T, Cerione RA, Zheng Y (1994) Cellular transformation and guanine nucleotide exchange activity are catalyzed by a common domain on the dbl oncogene product. J Biol Chem 269:62-65.

Hartwig JH (1992) Mechanisms of actin rearrangments mediating platelet activation. J Cell Biol 118:1421-1442.

Hartwig JH, Bokoch GM, Carpenter CL, Janmey PA, Taylor LA, Toker A, Stossel TP (1995) Thrombin receptor ligation and activated Rac uncap actin filaments barbed ends through phosphoinositide synthesis in permeabilized human platelets. Cell 82:643-653.

Heisterkamp N, Kaartinen V, van Soest S, Bokoch GM, Groffen J (1993) Human ABR encodes a protein with GAPrac activity and homology to the DBL nucleotide exchange factor domain. J Biol Chem 268:16903-16906.

Herrera R, Shivers BD (1994) Expression of alpha 1-chimerin (rac-1 GAP) alters the cytoskeletal and adhesive properties of fibroblasts. J Cell Biochem 56:582-591.

Hildebrand JD, Taylor JM, Parsons JT (1996) An SH3 domain-containing GTPase-activating protein for Rho and Cdc42 associates with focal adhesion kinase. Mol Cell Biol 16:3169-3178.

Homma Y, Emori Y (1995) A dual functional signal mediator showing RhoGAP and phospholipase C-delta stimulating activities. EMBO J 14:286-291.

Horiguchi Y, Senda T, Sugimoto N, Katahira J, Matsuda M (1995) *Bordetella bronchiseptica* dermonecrotizing toxin stimulates assembly of actin stress fibers and focal adhesions by modifying the small GTP-binding protein Rho. J Cell Sci 108:3243-3251.

Horii Y, Beeler JF, Sakaguchi K, Tachibana M, Miki T (1994) A novel oncogene, ost, encodes a guanine nucleotide exchange factor that potentially links Rho and Rac signaling pathways. EMBO J 13:4776-4786.

Hotchin NA, Hall A (1995) The assembly of integrin adhesion complexes requires both extracellular matrix and intracellular Rho/Rac GTPases. J Cell Biol 131:1857-1865.

Huttenlocher A, Sandborg RR, Horwitz AF (1995) Adhesion in cell migration. Curr Opin Cell Biol 7:697-706.

Ishizaki T, Maekawa M, Fujisawa K, Okawa K, Iwamatsu A, Fujita A, Watanabe N, Saito Y, Kakizuka A, Morii N, Narumiya S (1996) The small GTP-binding protein Rho binds to and activates a 160 kDa Ser/Thr protein kinase homologous to myotonic dystrophy kinase. EMBO J 15:1885-1893.

Johnson RP, Craig SW (1995) F-actin binding site masked by the intramolecular association of vinculin head and tail domains. Nature 373:261-264

Joneson T, McDonough M, Bar-Sagi D, Van Aelst L (1996) Rac regulation of actin polymerization and proliferation by a pathway distinct from Jun Kinase. Science 274:1374-1376.

Jullien-Flores V, Dorseuil O, Romero F, Letourneur F, Saragosti S, Berger R, Tavitian A, Gacon G, Camonis JH (1995) Bridging Ral GTPase to Rho pathways. RLIP76, a Ral effector with CDC42/Rac GTPase-activating protein activity. J Biol Chem 270:22473-22477.

Just I, Selzer J, Wilm M, von Eichel-Streiber C, Mann M, Aktories K (1995) Glucosylation of Rho proteins by Clostridium difficile toxin B. Nature 375:500-503.

Kaibuchi K, Mizuno T, Fujioka H, Yamamoto T, Kishi K, Fukumoto Y, Hori Y, Takai Y (1991) Molecular cloning of the cDNA for stimulatory GDP/GTP exchange protein for smg p21s (ras p21-like small GTP-binding proteins) and characterization of stimulatory GDP/GTP exchange protein. Mol Cell Biol 11:2873-2880.

Kenney D, Caims L, Remold-O'Donnell E, Peterson J, Rosen FS, Parkman R (1986) Morphological abnormalities in the lymphocytes of patients with the Wiskott-Aldrich syndrome. Blood 68:1329-1332.

Kimura K, Ito M, Amano M, Chihara K, Fukata Y, Nakafuku M, Yamamori B, Feng J, Nakano T, Okawa K, Iwamatsu A, Kaibuchi K (1996) Regulation of myosin phosphatase by Rho and Rho-associated kinase (Rho-kinase). Science 273:245-248.

Kolluri R, Fuchs Tolias K, Carpenter CL, Rosen FS, Kirchhausen T (1996) Direct interaction of the Wiskott-Aldrich syndrome protein with the GTPase Cdc42. Proc Natl Acad Sci USA 93:5615-5618.

Kozma R, Amhed S, Best A, Lim L (1995) The Ras-related protein CDC42Hs and bradykinin promote formation of peripheral actin microspikes and filopodia in Swiss 3T3 fibroblasts. Mol Cell Biol, 15:1942-1952.

Kumagai N, Morii N, Fujisawa K, Nemoto Y, Narumiya S (1993) ADP-ribosylation of rho p21 inhibits lysophosphatidic acid-induced protein tyrosine phosphorylation and phosphatidylinositol 3-kinase activation in cultured Swiss 3T3 cells. J Biol Chem 268:24535-24538.

Lamarche N, Tapon N, Stowers L, Burbelo PD, Aspenstrom P, Bridges T, Chant J, Hall A (1996) Rac and Cdc42 induce actin polymerization and G1 cell cycle progression independently of p65Pak and the JNK/SAPK MAP kinase cascade. Cell 87:519-529.

Lancaster CA, Taylor-Harris PM, Self AJ, Brill S, van Erp HE, Hall A (1994) Characterization of rhoGAP. A GTPase-activating protein for rho-related small GTPases. J Biol Chem 269:1137-1142.

Laudanna C, Campbell JJ, Butcher EC (1996) Role of Rho in chemoattractant-activated leukocyte adhesion through integrins. Science 271:981-983.

Leung T, How BE, Manser E, Lim L (1993) Germ cell beta-chimerin, a new GTPase-activating protein for p21rac, is specifically expressed during the acrosomal assembly stage in rat testis. J Biol Chem 268:3813-3816.

Leung T, Manser E, Tan L, Lim L (1995) A novel serine/threonine kinase binding the Ras-related RhoA GTPase which translocates the kinase to peripheral membranes. J Biol Chem 270:29051-29054.

Leung T, Chen XQ, Manser E, Lim L (1996) The p160 Rokα is a member of a kinase family and is involved in the reorganization of the cytoskeleton. Mol Cell Biol 16:5313-5327.

Luo L, Liao YJ, Jan LY, Jan YN (1994) Differential effects of the Rac GTPase on Purkinje cell axons and dendritic trunks and spines. Distinct morphogenetic functions of similar small GTPases: Drosophila Drac1 is involved in axonal outgrowth and myoblast fusion. Genes Dev 8:1787-1802.

Luo L, Hensch TK, Ackerman L, Barbel S, Jan LY, Jan YN (1996) Differential effects of the Rac GTPase on Purkinje cell axons and dendritic trunks and spines. Nature 379:837-840.

Machesky LM, Hall A (1996) Rho: a connection between membrane receptor signalling and the cytoskeleton. Trends Cell Biol 6:304-310.

Madaule P, Furuyashiki T, Reid T, Ishizaki T, Watanabe G, Morii, N, Narumiya S (1995) A novel partner for the GTP-bound forms of rho and rac. FEBS Lett 377:243-248.

Malcolm KC, Ross AH, Qui RG, Symons M, Exton JH (1994) Activation of rat liver phospholipase D by the small GTP-binding protein RhoA. J Biol Chem 269:25951-25954.

Manser E, Chong C, Zhao ZS, Leung T, Michael G, Hall C, Lim L (1995) Molecular cloning of a new member of the p21-Cdc42/Rac-activated kinase (PAK) family. J Biol Chem 270:25070-25078.

Manser E, Leung T, Salihuddin H, Tan L, Lim L (1993) A nonreceptor tyrosine kinase that inhibits the GTPase activity of p21Cdc42. Nature 363:364-367.

Manser E, Leung T, Salihuddin H, Zhao ZS, Lim L (1994) A brain serine/threonine protein kinase activated by Cdc42 and Rac1. Nature 367:40-46.

Martin GA, Bollag G, McCormick F, Abo A (1995) A novel serine kinase activated by Rac/Cdc42Hs-dependent autophosphorylation is related to PAK65 and Ste20. EMBO J 14:1970-1978.

Matsui T, Amano M, Yamamoto T, Chihara K, Nakafuku M, Ito M, Nakano T, Okawa K, Iwamatsu A, Kaibuchi K (1996) Rho-associated kinase, a novel serine/threonine kinase, as a putative target for small GTP binding protein Rho. EMBO J 15:2208-2216.

McGlade J, Brunkhorst B, Anderson D, Mbamalu G, Settleman J, Dedhar S, Rozakis-Adcock M, Chen LB, Pawson T (1993) The N-terminal region of GAP regulates cytoskeletal structure and cell adhesion. EMBO J 12:3073-3081.

Michiels F, Habets GGM, Stam JC, van der Kammen R, Collar JG (1995) A role for Rac in Tiam-1-induced membrane ruffling and invasion. Nature 375:338-340.

Mitchinson TJ, Cramer LP (1996) Actin-based cell motility and cell locomotion. Cell 84:371-379.

Molina IJ, Kenney DM, Rosen FS, Remold-O'Donnell E (1992) T cell lines characterize events in the pathogenesis of the Wiskott-Aldrich syndrome. J Exp Med 176:867-874.

Morii N, Teruuchi T, Tominaga T, Kumagai N, Kazaki S, Ushikubi F, Narumiya S (1992) A Rho gene product in human blood platelets. II Effects of the ADP-ribosylation by botulinum C3 ADP ribo-syltransferase on platelet aggregation. J Biol Chem 267:20921-20926.

Murphy AM, Montell DJ (1996) Cell type specific roles for Cdc42, Rac and Rho L in Drosophila oogenesis. J Cell Biol 133:617-630.

Nakagawa O, Fujisawa K, Ishizaki T, Saito Y, Nakao K, Narumiya S (1996) ROCK-I and ROCK-II, two isoforms of Rho-associated coiled-coil forming protein serine/threonine kinase in mice. FEBS Lett 392:189-193.

Narumiya S, Morii N, Sekine A, Kozaki S (1990) ADP-ribosylation of the rho/rac gene products by botulinum ADP-ribo-syltransferase: identity of the enzyme and effects on protein and cell functions. J Physiol 84:267-272.

Nobes CA, Hall A (1995) Rho, Rac and Cdc42 GTPases regulate the assembly of multimolecular focal complexes associated with actin stress fibers, lamellipodia and filopodia. Cell 81:53-62.

Nusrat A, Giry M, Turner JR, Colgan SP, Parkos CA, Carnes D, Lemichez E, Boquet P, Madara JL (1995) Rho protein regulates tight junctions and perijunctional actin organization in polarized epithelia. Proc Natl Acad Sci USA 92:10629-10633.

Ochs HD, Slichter SJ, Harker LA, Von Behrens WE, Clark RA, Wedgwood RJ (1980) The Wiskott-Aldrich syndrome: studies of lymphocytes, granulocytes, and platelets. Blood 55:243-252.

Ogha N, Kikuchi A, Ueda T, Yamamoto J, Takai Y (1989) Rabbit intestine contains a protein that inhibits the dissociation of GDP from and the subsequent binding of GTP to RhoB p20, a ras p21-like GTP-binding protein. Biochem Biophys Res Comm 163:1523-1533.

Pai EF, Kabsch W, Krengel U, Holmes KC, John J, Wittinghofer A (1989) Structure of the guanine-nucleotide-binding domain of the Ha-ras oncogene product p21 in the triphosphate conformation. Nature 341:209-214.

Parsons TJ (1996) Integrin-mediated signaling: regulation by protein tyrosine kinases and small GTP-binding proteins. Curr Opin Cell Biol 8:146-152.

Paterson HF, Self AJ, Garrett MD, Just I, Aktories K, Hall A (1990) Microinjection of recombinant p21rho induces rapid changes in cell morphology. J Cell Biol 111:1001-1007.

Peppelenbosch MP, Qiu RG, de Vries-Smits AM, Tertoolen LG, de Laat SW, McCormick F, Hall A, Symons MH, Bos JL (1995) Rac mediates growth factor-induced arachidonic acid release. Cell 81:849-856.

Philips MR, Pillinger MH (1995) Prenylcysteine-directed carboxyl methyltransferase activity in human neuthophil membranes. Methods Enzymol 256:49-63.

Prigmore E, Ahmed S, Best A, Kozma R, Manser E, Segal AW, Lim L (1995) A 68-kDa kinase and NADPH oxidase component p67phox are targets for Cdc42Hs and Rac1 in neutrophils. J Biol Chem 270:10717-10722.

Reid T, Furuyashiki T, Ishizaki T, Watanabe G, Watanabe N, Fujisawa K, Morii N, Madaule P, Narumiya S (1996) Rhotekin, a new putative target for Rho bearing homology to a serine/threonine kinase, PKN, and rhophilin in the rho-binding domain. J Biol Chem 271:13556-13560.

Reinhard J, Scheel AA, Diekmann D, Hall A, Ruppert C, Bahler M (1995) A novel type of myosin implicated in signalling by rho family GTPases. EMBO J 14:697-704.

Ren XD, Bokoch GM, Anderson RA, Jenkins GH, Schwartz MA (1996) Physical association of the small GTP-binding protein Rho with a phosphatidylinositol 4-phosphate 5-kinase in Swiss 3T3 cells. Mol Biol Cell 7:435-442.

Ridley AJ, Hall A (1992) The small GTP-binding protein Rho regulates the assembly of focal adhesions and actin stress fibers in response to growth factors. Cell 70:389-399.

Ridley AJ, Hall A (1994) Signal transduction pathways regulating Rho-mediated stress fiber formation: requirement for a tyrosine kinase. EMBO J 13:2600-2610.

Ridley AJ, Paterson HF, Johnston CL, Diekmann D, Hall A (1992) The small GTP-binding protein Rac regulates growth factor-induced membrane ruffling. Cell 70:401-410.

Ridley AJ, Self AJ, Kasmi F, Paterson HF, Hall A, Marshall CJ, Ellis C (1993) Rho family GTPase activating proteins p190, bcr, and rhoGAP show distinct specificities in vitro and in vivo. EMBO J 12:5151-5160.

Ron D, Zannini M, Lewis M, Wickner RB, Hunt LT, Graziani G, Tronick SR, Aaronson SA, Eva A (1991) A region of proto-dbl essential for its transforming activity shows sequence similarity to a yeast cell cycle gene, CDC24, and the human breakpoint cluster gene, bcr. New Biol 3:372-379.

Rubin EJ, Gill DM, Boquet P, Popoff MR (1988) Functional modification of a p21 kDa G-protein when ADP-ribosylated by exoenzyme C3 of Clostridium botulinum. Mol Cell Biol 8:418-426.

Scherle P, Behrens T, Staudt LM (1993) Ly-GDI, a GDP-dissociation inhibitor of the RhoA GTP-binding protein, is expressed preferentially in lymphocytes. Proc Natl Acad Sci USA 90:7568-7572.

Schmidt M, Rumenapp U, Bienek C, Keller J, von Eichel-Streiber C, Jakobs KH (1996) Inhibition of receptor signaling to phospholipase D by Clostridium difficile toxin B. Role of Rho proteins. J Biol Chem 271:2422-2426.

Seckl MJ, Morii N, Narumiya S, Rozengurt E (1995) Guanosine 5'-3-O-(thio)triphosphate stimulates tyrosine phosphorylation of p125FAK and paxillin in permeabilized Swiss 3T3 cells. Role of p21rho. J Biol Chem 270:6984-6990.

Self AJ Hall A (1995) Purification of recombinant Rho/Rac/G25K from Escherichia Coli. Methods Enzymol 256:3-10.

Self AJ, Paterson HF, Hall A (1993) Different structural organization of Ras and Rho effector domains. Oncogene 8:655-661.

Settleman J, Albright CF, Foster LC, Weinberg RA (1992) Association between GTPase activators for Rho and Ras families. Nature 359:153-154.

Small JV (1994) Lamellipodia architecture: actin filaments turnover and the lateral flow of actin filaments during motility. Seminars Cell Biol 5:157-163.

Stossel TP (1993) On the crawling of animal cells. Science 260:1086-1094.

Stowers L, Yelon D, Berg LJ, Chant J (1995) Regulation of the polarization of T cells toward antigen-presenting cells by Ras-related GTPase Cdc42. Proc Natl Acad Sci USA 92:5027-5031.

Sun HQ, Kwiatkowska K, Yin HL (1995) Actin monomer binding proteins. Curr Opin Cell Biol 7:102-110.

Symons M, Derry JM, Karlak B, Jiang S, Lemahieu V, Mccormick F, Francke U, Abo A (1996) Wiskott-Aldrich syndrome protein, a novel effector for the GTPase CDC42Hs, is implicated in actin polymerization. Cell 84:723-734.

Takai Y, Sasaki T, Tanaka K, Nakanishi H (1995) Rho as a regulator of the cytoskeleton. Trends Biochem Sci 20:227-231.

Tan EC, Leung T, Manser E, Lim L (1993) The human active breakpoint cluster region-related gene encodes a brain protein with homology to guanine nucleotide exchange proteins and GTPase-activating proteins. J Biol Chem 268:27291-27298.

Tolias KF, Cantley LC, Carpenter CL (1995) Rho family GTPases bind to phosphoinositide kinases. J Biol Chem 270:17656-17659.

Tominaga T, Sugie K, Hirata M, Morii N, Fukata J, Uchida A, Imura H, Narumiya S (1993) Inhibition of PMA-induced, LFA-1-dependent lymphocyte aggregation by ADP ribosylation of the small molecular weight GTP binding protein Rho. J Cell Biol 120:1529-1537.

Udagawa T, McIntyre BW (1996) ADP-ribosylation of the G protein Rho inhibits integrin regulation of tumor cell growth. J Biol Chem 271:12542-12548.

Van Aelst L, Joneson T, Bar-Sagi D (1996) Identification of a novel Rac1-interacting protein involved in membrane ruffling. EMBO J 15:3778-3786.

Watanabe G, Saito Y, Madaule P, Ishizaki T, Fujisawa K, Morii N, Mukai H, Ono Y, Kakizuka A, Narumiya S (1996) Protein kinase N (PKN) and PKN-related protein rhophilin as targets of small GTPase Rho. Science 271:645-8.

Weekes J, Barry ST, Critchley DR (1996) Acidic phospholipids inhibit the intramolecular association between the N-terminal and C-terminal regions of vinculin, exposing actin-binding and protein kinase C phosphorylation sites. Biochem J 314:827-832.

Zhang J, King WG, Dillon S, Hall A, Feig L, Rittenhouse SE (1993) Activation of platelet phosphatidylinositide 3-kinase requires the small GTP-binding protein Rho. J Biol Chem, 268:22251-22254.

Zhang J, Zhang J, Benovic JL, Sugai M, Wetzker R, Gout I, Rittenhouse SE (1995) Sequestration of a G-protein beta gamma subunit or ADP-ribosylation of Rho can inhibit thrombin-induced activation of platelet phosphoinositide 3-kinases. J Biol Chem 270:6589-6594.

Zhang Z, Vuori K, Wang HG, Reed JC, Ruoslahti E (1996) Integrin activation by R-Ras. Cell 85:61-69.

Zheng Y, Bagrodia S, Cerione RA (1994) Activation of phosphoinositide 3-kinase activity by Cdc42Hs binding to p85. J Biol Chem 269:18727-18730.

Zheng Y, Olson MF, Hall A, Cerione RA, Toksoz D (1995) Direct involvement of the small GTP-binding protein Rho in lbc oncogene function. J Biol Chem 270:9031-9034.

CHAPTER 7

Regulation of Gene Expression

PATHWAYS LEADING TO GENE REGULATION

In most cases the end point of the receptor-induced signaling cascade is to regulate gene expression and thus to modify cellular metabolisms, differentiation or proliferation. The signal from the transmembrane receptors should proceed through the cytoplasm to the nucleus and ultimately modify the activity of DNA-binding proteins that regulate transcription of specific DNA sequences. Four major pathways are known that lead to regulation of gene expression in response to extracellular stimuli.

(1) The cAMP/protein kinase A pathway that regulates transcription of the genes containing the DNA sequence CRE (c-AMP Responsive Element) in their promoters (Fig. 7.1).

cAMP, generated by G protein-stimulated adenylate cyclase, binds to the regulatory subunit of the protein kinase A (PKA) allowing the catalytic subunit to translocate to the nucleus and phosphorylate CREB, a protein that binds to the CRE sequence in the DNA and activates transcription (Lalli and Sassone-Corsi, 1994).

(2) The monomeric GTPases/MAPK pathway (see chapter 5) that regulates transcription of gene promoters containing the SRE (Serum Responsive Element) or the TRE (TPA responsive element) sequences (Fig. 7.1).

Upon activation, Erk1 and Erk2 MAPK translocate to the nucleus and phosphorylate proteins of the DNA binding factor TCF (Ternary Complex Factor) that include Elk-1 and SAP-1 (Hill and Treisman, 1995; Su and Karin, 1996). TCF forms a complex with SRF (Serum response factor) and together they regulate transcription of those genes containing the SRE in their promoters such as the c-fos promoter (Hill and Treisman, 1995; Su and Karin, 1996). The SRF activity is also regulated, but via a different pathway involving RhoA, Rac1 and Cdc42, monomeric GTPases belonging to the Rho subfamily (Vojtek and Cooper, 1995). A third pathway involving Rac1 and Cdc42 is responsible for the activation of the Jnk1 and Jnk2 MAPK that regulates the transcription factors c-Jun and ATF-2 (Hill et al 1995; Minden et al 1995; Coso et al 1995; Su and Karin, 1996). c-Jun and ATF-2 bind to the TRE present in the c-Jun gene promoter as well as in the promoter sequences of several other genes. c-Jun can, thus, transactivate its own gene.

(3) The Jak-Stat pathway that leads to activation of the transcription complexes Stats (Signal transducer and activator of transcription) (Fig. 7.1).

Many cytokine receptors lacking intrinsic catalytic activity activate soluble tyrosine kinases of the Janus family (Jak1, Jak2, Jak3 and Tyk2). These activated kinases can

Signal Transduction by Integrins, by Paola Defilippi, Angela Gismondi, Angela Santoni and Guido Tarone. © 1997 Landes Bioscience.

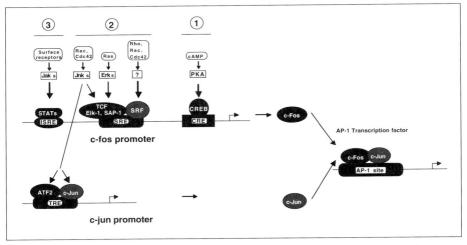

Fig 7.1. The c-fos and the c-jun genes are among the immediate early genes induced by growth factor response and contain several regulatory sequence in their promoters: 1) the CRE sequences regulated by the cAMP and PKA; 2) the SRE sequence regulated by the small GTPases Ras, Rho, Rac, Cdc42 and MAPKs Erk1 and 2; 3) the TRE sequence regulated by Rac and Cdc42 and MAPKs of the Jnk family; and 4) the ISRE sequence regulated by membrane receptors via the Jak kinases. Activation of c-fos and c-jun gene transcription leads to increased levels of the corresponding proteins c-Fos and c-Jun that, after dimerization, form the transcription factor AP-1 that can transactivate a second wave of gene expression.

phosphorylate cytoplasmic proteins belonging to the family of transcription factors called Stats (Fig. 7.1). The Stats consist of a number of proteins, including p84, p91 and p113, that upon tyrosine phosphorylation, associate with other subunits to form a transcriptional complex active on the ISRE (Interferon-Stimulated Response Element). Several different Stat factors have been identified and are numbered from 1 to 6 (Schindler and Darnell, 1995).

(4) The pathway that leads to activation of the transcription factor NF-κB (Fig. 7.2).

NF-κB belongs to the Rel family of transcription factors. Its activity is regulated through the interaction with IκB. Inactivation of the IκB subunit by phosphorylation and proteolysis leads to translocation of the NF-κB to the nucleus and transcriptional activation of the genes bearing κB sites in their promoters and enhancers (Fig. 7.2). Signals leading to activation of NF-κB may involve, among others, PKC activation, sphingomyelin degradation and activation of Raf. However, the precise pathway(s) have

not yet been elucidated (Siebenlist et al 1994).

INTEGRIN-DEPENDENT REGULATION OF GENE EXPRESSION

Evidence that cell adhesion can affect gene expression was provided by experiments in the late seventies showing that DNA synthesis is strictly linked to cell spreading on a solid substratum (Folkman and Moscona, 1978). Since then, several studies in different cellular systems have expanded this observation and shown that both proliferation (see chapter 8) and differentiation (Adams and Watt, 1989; Streuli et al 1991; Menko and Boettiger, 1987; West et al 1979) depend upon cell adhesion and integrin function. Since proliferation and differentiation require profound changes in cellular functions, it is safe to conclude that cell-matrix interactions can affect gene expression. However, direct demonstration for matrix-dependent gene regulation has been reported in a few cases (see below) and only

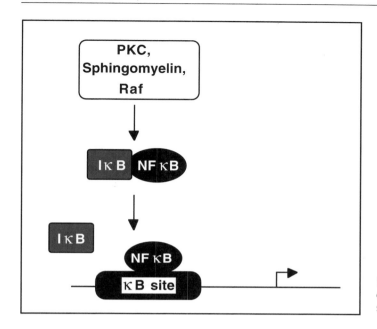

Fig. 7.2. Activation of the NF-κB transcription factor.

recently, specific promoter/enhancer sequences and transcription factors regulated by matrix-derived signals have been identified (Schmidhauser et al 1990, 1992, 1994; DiPersio et al 1991; Haskill et al 1991; Liu et al 1991; Yamada et al 1991; Rana et al 1994; Fan et al 1995; Tremble et al 1995).

The integrin derived signals that control cell proliferation are discussed in chapter 8. Here we review the data concerning the broad general effect of cell adhesion on the regulation of gene expression. We will discuss different cellular systems separately and in particular the effect of extracellular matrix on differentiation and gene expression in epithelial cells, leukocytes and mesenchymal cells.

DIFFERENTIATION OF EPITHELIAL CELLS IS AFFECTED BY CELL-MATRIX INTERACTION

Epithelial cells adhere to basement membranes consisting of laminin, nidogen, collagen IV and proteoglycans. During wound healing and tissue remodeling, epithelial cells may interact with stromal components such as fibronectin or collagen I.

Fibronectin inhibits terminal differentiation of human keratinocytes (Adams and Watt, 1989; Watt et al 1993). When cultured in suspension, keratinocytes stop proliferating and undergo terminal differentiation, suggesting that interaction with basal membrane components is important for keratinocyte proliferation and maintenance of the undifferentiated state. If fibronectin is added to the suspension culture, keratinocytes still withdraw from the cell cycle, but terminal differentiation is inhibited (Fig. 7.3). This response may provide an explanation for the fact that only suprabasal keratinocytes, that are no longer in contact with the extracellular matrix, undergo terminal differentiation. This effect may also represent a mechanism to suppress keratinocyte differentiation during epidermal wound healing. The action of fibronectin is mediated by the α5β1 integrin receptor, but the signaling mechanisms involved in this process have not been investigated.

Basement membrane components exert a profound effect on the expression of the milk protein β casein in mammary gland

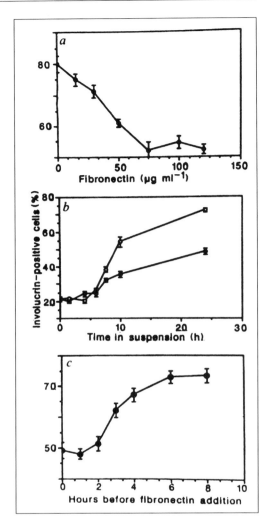

Fig. 7.3. Fibronectin inhibition of keratinocyte terminal differentiation. (a) Different concentrations of fibronectin were added at the time of suspending cells and the percentage of involucrin-positive cells measured 24 h later. (b) Percentage of cells expressing involucrin was measured after different times in suspension in the presence of bovine serum albumin (75 μg/ml) (▯) or fibronectin (▮). (c) Fibronectin (75 μg/ml) was added at different times after plating and the cells were collected after 24 h in suspension. With permission from Adams JC and Watt FM. Nature 1989; 340:307-309. Copyright 1989 Macmillan Magazines Limited.

cells. Synthesis of β casein requires the combined action of basement membrane components (Matrigel) and prolactin (Lee et al 1985; Li et al 1987). The specific basement membrane molecule involved in β casein expression has not been identified but laminin and heparin sulfate proteoglycan seem to play a major role (Li et al 1987). The matrix signals do not require polarization of the cells and are blocked by β1 integrin antibodies (Streuli et al 1991; Roskelley et al 1994). Further evidence that matrix proteins can affect β casein expression also comes from the observation that the decrease in β casein expression occur-

ring during involution of mammary epithelium after lactation, correlates with degradation of matrix proteins by the 72 kDa gelatinase, stromelysin and tissue plasminogen activator (Talhouk et al 1992; Sympson et al 1994). Basement membrane proteins, but not fibronectin or collagen, also prevent apoptosis of mammary epithelial cells presumably through specific integrin heterodimer signaling (Boudreau et al 1995). The increased expression of the β casein in response to extracellular matrix is due to transcriptional regulation and a 160 bp sequence with transcriptional enhancer activity (BCE1) has been identified

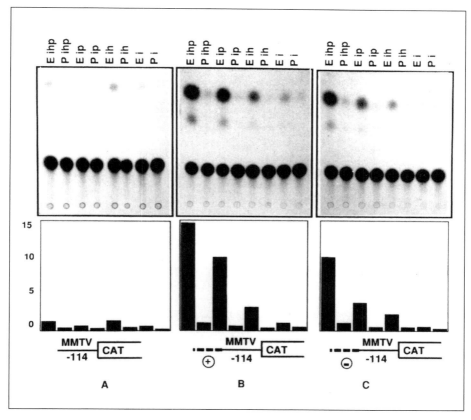

Fig. 7.4. Effect of the BCE1 (β casein enhancer (1) sequence on the transcription of a CAT reporter gene under a viral promoter. To test the enhancer activity, BEC1 sequences (dashed lines) were linked in a positive (B) or negative orientation (C) to a truncated form (-114 bp) of the mouse mammary tumor virus (MMTV) promoter, that per se is unable to drive the expression of the reporter gene CAT (A). CID 9 cells were stably transfected and cultured for 6 days in defined medium on either EHS basement membrane proteins (E) or tissue culture plastic (P) in the presence of insulin (i), hydrocortisone (h), prolactin (p) or combination of these. Two individual transfection sets varied up to 19% in total activity whereas the regulation patterns remained constant. Ten μg of protein lysate were used for each of the CAT assays. Activity: cpm/μg^{-1}/min^{-1}. Note that the combination of insulin, hydrocortisone and prolactin is ineffective to induce maximal CAT reporter activity in the absence of the extracellular matrix (first two lanes panel B and C). From Schmidhauser C et al Mol Biol Cell 1992; 3:699-709, with permission of the American Society for Cell Biology.

in the 5' flanking region of the casein gene (Schmidhauser et al 1990 and 1992). The BCE1 sequence contains the responsive elements for both prolactin and basement membrane-dependent regulation (Fig. 7.4), but specific transcription factors or signaling pathways involved have not yet been identified.

β-lactoglobulin synthesis is also controlled by the combination of prolactin and matrix stimuli (Streuli et al 1995). Prolactin signaling pathways lead to tyrosine phosphorylation of Stat5 via Jak2. Phosphorylated Stat5 translocates to the nucleus and binds to a site in the β-lactoglobulin gene promoter activating transcription. The ability of Stat5 to bind its cognate DNA sequence in the β-lactoglobulin gene promoter and drive transcription, however, requires interaction of the cells with the basement membrane in addition to the prolactin stimulus (Streuli et al 1995). Thus, Stat5 transcription factor is a target of the matrix dependent control on gene expression.

Cell-matrix interaction also exerts a profound effect on gene expression in hepatocytes. Albumin synthesis is strongly stimulated by basement membrane extracts as well as by laminin but not by heparin sulfate proteoglycan or collagen IV (Bissell et al 1990; Caron 1990). This seems to be mediated by two liver transcription factors, HFN3 and eH-TF, whose synthesis and DNA binding activity is stimulated by extracellular matrix (Liu et al 1991; DiPersio et al 1991). Integrin binding to Arg-Gly-Asp peptides or fibronectin, moreover, induces expression of JunB and Ras (Hansen et al 1994).

CELL-MATRIX INTERACTION REGULATES THE IMMUNE AND INFLAMMATORY RESPONSES IN LEUKOCYTES

Upon tissue injury, infection and remodeling, leukocytes should extravasate into the tissue stroma. During this process they start producing a variety of cytokines that trigger the inflammatory response.

Adhesion of blood mononuclear cells to solid surfaces is, per se, a stimulus that triggers expression of several cytokine genes. In particular high steady-state levels of transcripts for IL-1β, TNFα, CSF-1, PDGFβ, IL-8, superoxide dismutase and IL-1ra (IL-1 receptor antagonist) accumulate in response to adhesion to matrix proteins or plastic dishes (Fig. 7.5) (Eierman et al 1989; Shaw et al 1990; Sporn et al 1990; Pacifici et al 1992; Yurochko et al 1992; Hershkowiz et al 1993; Miyake et al 1993). At the same time, other genes, including c-fms, lysozyme (Fig. 7.5) and CD-4, are downregulated (Eierman et al 1989), indicating that monocyte-matrix adhesion exerts a selective action on gene expression in these cells. Some

of these genes, such as TNFα, are induced quite rapidly (30 min), while others, such as CSF-1, are induced later (90-120 min) after monocyte-matrix interaction (Eierman et al 1989). By differential hybridization of cDNA libraries from adherent and nonadherent monocytes, several additional adhesion-regulated genes have been isolated and named MAD (Monocyte Adhesion) genes (Sporn et al 1990); the identity of many of these genes is still unknown. In addition to cytokines and inflammatory mediators, several DNA-binding proteins with transcriptional activity are also induced upon monocyte-matrix adhesion. These include c-Fos (Fig. 7.5), c-Jun, NF-κB, IκBα, and A20 (Haskill et al 1988 and 1991; Shaw et al 1990; Sporn et al 1990; Lofquist et al 1995; Thieblemont et al 1995). Interestingly, the promoter sequences of all cytokine genes that are upregulated, but not of those that are downregulated, by monocyte-matrix adhesion contain NF-κB sites (Juliano and Haskill 1993), suggesting that NF-κB plays a major role in the upregulation of gene expression in response to adhesion. This hypothesis is also supported by a study on the tissue factor gene (TF), another gene induced in adherent monocytes (Fan et al 1995). The promoter region of the TF gene contains an integrin responsive element with two AP-1 sites and a single κB-like site. Using THP-1 monoblastic leukemia cells as monocyte model system, it was shown that both sites are required for the integrin-dependent response. However, triggering of β1 integrins in these cells leads to nuclear translocation of the c-Rel/p65 heterodimer (belonging to the NF-κB family) and binding to the κB-like site but does not stimulate DNA-binding activity of the AP-1 complex (Fan et al 1995). Thus, although both factors are required for transcriptional activity, the NF-κB-like activity is responsible for the integrin-induced increase of transcription. Activation of NF-κB has also been reported in fibroblasts and smooth muscle cells adhering to fibronectin (Qwarnstrom et al 1994), suggesting that involvement of this transcription factor in the integrin-in-

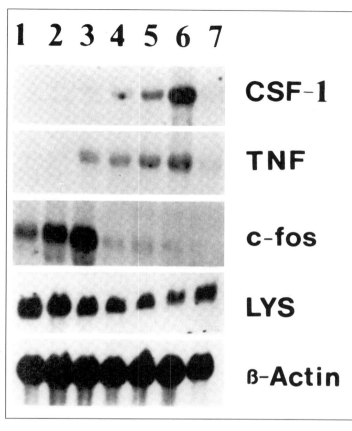

Fig. 7.5. Time course of steady-state mRNA of CSF-1, TNF-α, c-fos and lysozyme during adhesion in serum-free conditions. Density isolated monocytes were adhered to polystyrene tissue culture dishes or cultured nonadherently in polypropylene tubes for the indicated time periods. Total RNA was isolated and mRNA levels analyzed by Northern blot hybridization. Representative Northern analysis: 1, 0 min control; 2, 20 min adherence; 3, 40 min adherence; 4, 90 min adherence; 5, 2 h adherence; 6, 4 h adherence; 7, 4 h nonadherent control. Note that mRNA level of CSF-1 and TNF-α increase, while lysozyme transcript decreases, starting from 40 min of adherence. c-fos mRNA rapidly rises in the first 20 min of adhesion and than decreases to very low level. With permission from Eierman DF et al J Immunol 1989; 142:1970-1976. Copyright 1989. The American Association of Immunologists.

duced gene expression is not restricted to immune cells.

The likely role of the NF-κB-like factors in the upregulation of gene transcription in adherent monocytes is in apparent contradiction with the increased levels of I-κBα transcript detected in these cells (Haskill et al 1991). I-κBα, in fact, inhibits NF-κB transcriptional activity by binding to it and preventing its nuclear translocation. However, detailed analysis of the I-κBα regulation during monocyte adhesion, indicated that the I-Bα protein level drops within 5 min of monocyte adhesion and strongly increases in the following 20 min (Lofquist et al 1995).

This, in principle, allows a rapid and transient activation of NF-κB that may account for the increased transcription of the adhesion-induced genes. Induction of I-κBα in adherent monocytes is thus likely to represent a negative feedback loop to downregulate NF-κB-dependent transcription triggered in the initial phase of adhesion.

Different matrix proteins trigger selective patterns of gene expression (Eierman et al 1989; Sporn et al 1990). Both collagen and fibronectin induce CSF-1 gene, while only collagen induces TNFα; laminin, on the other hand, is a relatively inefficient stimulus for gene expression. The monocyte

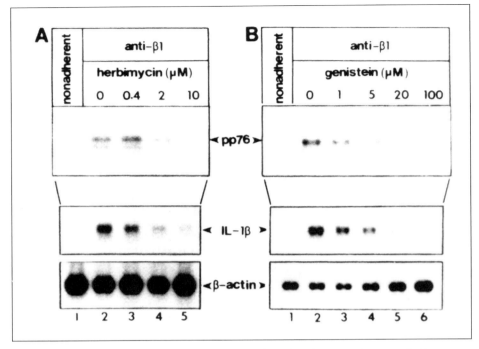

Fig. 7.6. Inhibition of integrin-induced tyrosine phosphorylation and IL1-β mRNA expression by herbimycin and genistein. Human monocytes were pretreated with various concentrations of herbimycin for 1 h at 37°C (A) or genistein for 15 min at 37°C (B). The cells were incubated with RPMI 1640 medium alone or medium containing anti-β1 integrin antibody TS2/16 (2 μg/ml) for 45 min on ice. The cells were washed and incubated with RPMI 1640 medium in the presence of inhibitors at 37°C. For western blotting, cell lysates harvested after 15 min incubation were probed with phosphotyrosine antibodies (2 x 10^5 cell/lane). For northern blotting, total cellular RNA isolated after 1 h incubation was probed with IL-1β or β-actin probes (3 μg total RNA/lane). Reproduced from Lin TH et al J Cell Biol 1994; 126:1585-1593, by copyright permission of The Rockefell University Press.

response to matrix proteins is mediated by β1 integrins as shown by the fact that antibody cross linking of β1, but not of β2 integrins, mimics gene transcription induced by matrix proteins (Yurochko et al 1992). The signaling pathway leading to transcriptional activation in monocytes involves activation of cytoplasmic tyrosine kinases, since specific inhibitors, such as genistein and herbimycin A, can block IL-1β message induction (Lin et al 1994) (Fig. 7.6). Monocytes do not express p125Fak, but tyrosine phosphorylation of a 70 kDa protein, probably corresponding to Syk, is triggered by β1 integrin clustering in these cells, suggesting that p72Syk may be part of the trans-

duction pathway to the nucleus (Fig. 7.6). Using the THP-1 monoblastic leukemia cell line as monocyte model system, it was shown that adhesion to fibronectin leads to increased tyrosine phosphorylation of two cytosolic tyrosine kinases, namely p125Fak and Syk, to translocation of NF-kB to the nucleus and increased levels of IL-1β mRNA. As shown for monocytes, in THP-1 cells tyrosine kinase inhibitors also block integrin-mediated tyrosine phosphorylation and increased IL-1β message, indicating a causal relationship between these two events (Lin et al 1995). β1 integrin clustering in suspended THP-1 cells induces tyrosine phosphorylation of Syk, but not of p125Fak, and

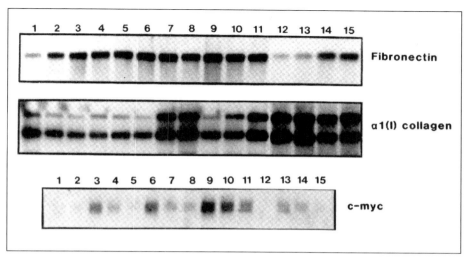

Fig. 7.7. Pro-α1(I)-collagen mRNA levels are induced by reattachment of suspension-arrested 3T3 cells. Cells were stimulated with serum while suspended or adherent or reattached in 0.4% or 10% serum. Total RNA was isolated from cells subjected to the following manipulation: asynchronous, subconfluent cells (lane 1); cells suspended for 48 h (lane 2); suspended cells stimulated in suspension for 2, 6 or 18 h (lanes 3-5) by addition of 10% serum; suspended cells reattached in 0.4% serum for 2, 6 or 18 h (lanes 6-8); suspended cells reattached in 10% serum for 2, 6 or 18 h (lanes 9-11); postconfluent cells (lane 12); postconfluent cells stimulated by addition of 10% serum for 2, 6 or 18 h (lanes 13-15). Note that adhesion is required to stimulate pro-α1(I)-collagen mRNA (compare lanes 3-5 with 6-8). With permission from Dhawan J and Farmer SR J Biol Chem 1990; 265:9015-9021.

increases IL-1β mRNA. This indicates that the tyrosine kinase Syk, but not p125Fak, may be a crucial component in the integrin-mediated pathway leading to gene regulation in monocytes (Lin et al 1995).

CELL ADHESION REGULATES MATRIX PROTEIN SYNTHESIS

When 3T3 fibroblasts are plated on culture dishes, type I collagen synthesis is strongly stimulated as compared with nonadherent cells. Attachment of suspended cells in very low serum (0.4%) causes a 7-fold induction of collagen mRNA levels (Fig. 7.7) and a greater than 20-fold rise in the rate of procollagen synthesis (Dhawan and Farmer 1990). This is not related to the growth stimulation of the cells and is due both to increased transcription, as tested by activation of a rat α1(I) promoter-chloramphenicol acetyltransferase reporter

gene construct, and to increased message stability (Dhawan et al 1991).

The mechanisms by which adhesion can trigger synthesis of matrix proteins are not known. However, the involvement of TGF-β1 has been suggested. In fact, cell adhesion can also regulate the transcription and synthesis of TGF-β1 (Streuli et al 1993), a cytokine known to upregulate matrix protein as well as integrin synthesis (Ignotz et al 1987; Ignotz and Massague 1987). In mammary epithelial cells the level of TGF-β1 expression is high in cells adherent on plastic but is strongly downregulated when cells are cultured on a reconstituted basement membrane matrix. Analysis of reporter gene expression under the control of the TGF-β1 promoter indicates that induction is due to transcriptional activation (Streuli et al 1993). Whether this property applies to other cellular systems has not been investigated, but these data raise the possibility that release of this cytokine may be responsible for the regulation of matrix protein synthesis in adherent cells. Upregulation of matrix protein synthesis in cells on plastic dishes may, thus, be secondary to TGF-β1 release.

Matrix protein synthesis in response to cell adhesion is regulated by the nature of the substratum. In fact, as mentioned above for TGF-β1, synthesis of laminin, type IV collagen and fibronectin mRNAs are high in mammary epithelial cells cultured on plastic surfaces but decreases when cells are grown on collagen type I coated dishes or on reconstituted basement membrane. This suggests that deposited matrix may send a negative signal that downregulates synthesis of new matrix components as well as that of TGF-β1 (Streuli and Bissell, 1990; Streuli et al 1993). This hypothesis is also supported by the observation that a significant downregulation of type I collagen synthesis is observed in fibroblasts grown in tridimensional collagen gels as compared with monolayer cultures (Mauch et al 1988). Downregulation of collagen type I in this system is controlled via the α1β1 integrin receptor (Langholz et al 1995; Riikonen et al 1995) both at transcriptional and post transcriptional levels (Eckes et al 1993).

CELL-MATRIX INTERACTIONS REGULATE SYNTHESIS OF MATRIX-DEGRADING ENZYMES

When grown in tridimensional matrix of type I collagen, fibroblasts release increased amounts of the interstitial collagenase MMP-1 (Unemori and Werb, 1986; Mauch et al 1989). Induction of MMP-1 occurs at transcriptional level (Mauch et al 1989), is controlled by the binding of collagen to the α2β1 integrin (Langholz et al 1995; Riikonen et al 1995) and requires activation of tyrosine kinases as shown by the ability of the specific inhibitor genistein to interfere with this response (Langholz et al 1995).

Synthesis of matrix proteinases, such as stromelysin, interstitial collagenase (MMP-1), the 72 kDa gelatinase (MMP-2), the 92 kDa gelatinase B and elastase, in response to cell-matrix interaction has also been demonstrated in synovial fibroblasts (Werb et al 1989; Tremble et al 1995), leukocytes (Xie et al 1993; Romanic and Madri, 1994), keratinocytes (Sudbeck et al 1994) and chondrocytes (Arner and Tortorella, 1995).

Increased synthesis and secretion of stromelysin and MMP-1 collagenase occurs upon synovial fibroblast adhesion to Arg-Gly-Asp-containing oligopeptides or to the 120 kDa fibronectin fragment containing the Arg-Gly-Asp cell binding site (Werb et al 1989). Antibodies to α5β1 integrin can mimic this signal indicating that integrins are responsible for induction of these metalloproteinase genes. Interestingly, intact fibronectin molecule does not induce metalloproteinase synthesis suggesting that domains outside the 120 kDa α5β1-binding fragment may suppress this function. Analysis of different fibronectin fragments indicated that the IIICS region of the molecule, containing the binding site for the α4β1 integrin, suppresses metalloproteinase expression induced by the 120 kDa fragment (Huhtala et al 1995) (Fig. 7.8). Thus α5β1 and α4β1 integrins generate an opposite response to fibronectin with respect to matrix degrading protease production. The fact that intact fibronectin does not trigger collagenase production suggests that α4β1 dominates over α5β1. It is possible that α4β1 represents the major functionally relevant fibronectin receptor in the cell studied or, alternatively, that α4β1 signaling interferes with the inductive signal from α5β1.

Mixed substrate of fibronectin and tenascin also stimulates metalloproteinase synthesis (Tremble et al 1994). Synovial fibroblasts plated on this substrate upregulate synthesis of four genes: MMP-1 collagenase, stromelysin, the 92 kDa gelatinase and c-fos. Tenascin does not induce the expression of collagenase in cells plated on substrates of type I collagen or vitronectin. Moreover, soluble tenascin added to cells adhering to a fibronectin substrate has no effect, sug-

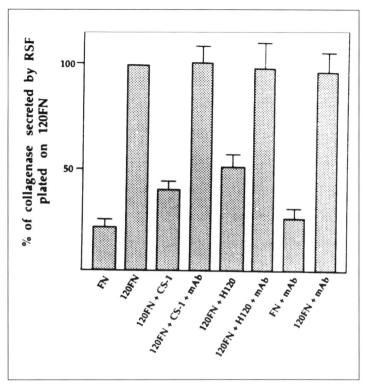

Fig. 7.8. Anti-α4 integrin monoclonal antibody blocks the ability of CS-1 and H120 to suppress induction by 120 kDa fibronectin fragment of collagenase expression in rabbit synovial fibroblasts (RSF). The following fibronectin fragments were used: 120FN, consisting of the central fibronectin domain with the α5β1-binding Arg-Gly-Asp site, but lacking the α4β1-binding IIICS region; H120, consisting of the heparin binding domain and the IIICS region, but lacking the Arg-Gly-Asp site; and CS-1, a 25 amino acid residue of the IIICS region corresponding to the α4β1 binding site. RSF were treated with anti-α4 integrin monoclonal antibody (5 µg/ml) for 30 min before plating them in wells coated with various substrates. RSF with or without anti-α4 treatment were plated on FN, 120FN, 120FN + CS-1 and 120FN + H120. Conditioned medium was analyzed by SDS-PAGE and immunoblotting with anticollagenase monoclonal antibodies. Collagenase expression was quantified by densitometry from six separate experiments, each in duplicate, and is shown as mean SEM. From Huhtala P et al J Cell Biol 1995; 129:867-879 by copyright permission of The Rockefeller University Press.

gesting that only mixed substrates are active. Collagenase increases within 4 h of cell plating on a fibronectin/tenascin substrate and exhibits kinetics similar to those for induction of collagenase gene expression by signaling through β1 integrins. Since tenascin is expressed in the matrix only in specific conditions, such as during mesenchymal-epithelial transition and in the stroma surrounding tumors (Mackie et al 1987; Inaguma et al 1988), it may be hypothesized that expression of tenascin regulates the expression of genes involved in cell migration, tissue remodeling and metastatic invasion of tumor cells.

α5β1-dependent induction of collagenase occurs at transcriptional level. Analysis of the minimal sequence responsive to integrin stimulation in the human collagenase gene promoter identified a region from -139/-67 (relative to the transcription start site) containing AP-1 and PEA3 sites (Tremble et al 1995). The AP-1 site seems to play an important role in the integrin-induced transcriptional activation of this gene. In fact, α5β1 integrin engagement in these cells also induces nuclear translocation of the c-Fos and c-Jun proteins and increased c-fos message. These events precede induction of the collagenase gene. A Jun/Jun homodimer or a Jun/Fos heterodimer forms the transcription factor that binds to the AP-1 site (see Fig. 7.1) and may thus regulate the integrin-induced collagenase gene expression.

Leukocytes regulate matrix degrading proteases release in response to adhesion. Adherent monocytes secrete increased amounts of elastase upon exposure to fibronectin fragments. Whether this is due to increased synthesis or secretion of internal stores has not been addressed (Xie et al 1993). T lymphocytes synthesize and release increased levels of the MMP-2 gelatinase upon adhesion to endothelial cells. This response is mediated by α4β1 integrin binding to the counter receptor VCAM-1 that is expressed on activated endothelial cells. Release of MMP-2 by T cells is important for their transendothelial migration as TIMP-2, an inhibitor of the gelatinase, partially prevents the migratory process (Romanic and Madri, 1994).

Collagenase induction has also been observed in keratinocytes. In this system, induction occurs when cells are plated on collagen type I, while basement membrane proteins are ineffective (Sudbeck et al 1994). This in vitro system mimics the in vivo situation during wound healing where migrating keratinocytes in contact with the dermal matrix, but not basal keratinocytes in the epidermis, express interstitial collagenase (Saarialho-Kere et al 1992 and 1993). In this system, the activity of integrin antibodies was not tested, but it is likely that these receptors are involved. Tyrosine kinases and protein kinase C are involved in the signaling pathway of collagen-mediated collagenase synthesis in keratinocytes as indicated by the use of specific inhibitors (Sudbeck et al 1994).

In addition to matrix degrading metalloproteinases, induction of the serine protease tPA and uPA (tissue-type and urokinase-type plasminogen activator, respectively) are induced in differentiated mouse teratocarcinoma cells by β1 integrin engagement or by disruption of the actin cytoskeleton with cytochalasin B (Snyder et al 1992).

The data discussed above indicate that a major effect on gene expression exerted by the interaction of cells with the extracellular matrix is the downregulation of matrix protein synthesis and upregulation of matrix-degrading protease synthesis. This control loop may be functionally relevant in different biological responses. The ability of collagen matrix to downregulate collagen synthesis may represent an important mechanism during tissue organization and repair. In fact, this allows the cells to sense the extracellular environment and to stop collagen synthesis when appropriate level of matrix deposition has been reached. At the same time, induction of collagenase production allows cells to re-elaborate the deposited matrix, a process occurring during wound healing (Grinnell 1994). The ability of matrix proteins or their degradation products to induce protease production may also be relevant to pathological situations. The ability of tumor cells to invade stromal tissue and propagate into the organism as metastasis can be increased by collagenase production. Increased levels of MMP-2 gelatinase are secreted by melanoma cells in response to αVβ3 or α5β1 integrin stimulation and this event correlates with increased invasiveness in vitro (Seftor et al 1992 and 1993). The α2β1 integrin, the collagen receptor mediating this response (Langholz et al 1995; Riikonen et al 1995), is expressed in highly aggressive melanoma cells (Klein et al 1991a and 1991b; Santala et al 1994) and forced expression of this receptor shifts noninvasive RD rhabdomyosarcoma cells into metastatic cells in vivo (Chan et al 1991). Moreover, the ability of fibronectin degradation products and of mixed tenascin/fibronectin matrices to trigger matrix proteases synthesis in synovial fibroblasts may be relevant in joint inflammatory pathologies.

CELL ADHESION REGULATES INTEGRIN EXPRESSION

Adhesion to extracellular matrix also affects integrin subunit synthesis, but contradictory results have been reported in different cellular systems.

The α2β1 collagen receptor is up-regulated both at transcriptional and translational level upon fibroblast interaction with tridimensional collagen gel (Klein et al 1991a). Chen and coworkers (1992), on the other hand, have shown that the α2, α4 and αV integrin subunits are markedly decreased at mRNA and protein levels, in MG63 osteosarcoma cells upon adhesion to culture dishes. Not all integrin subunits, however, are regulated as shown by unaltered α5 expression. Other reports show that cell surface integrins expression is controlled by adhesion. In fact, plasma membrane integrins decrease 5-50-fold, depending on the cell line examined upon detachment from the substratum, due to increased internalization and degradation of the molecules, but not to decreased synthesis (Dalton et al 1995). Interestingly, ligation of α5β1 receptor by fibronectin also leads to increased surface expression of the collagen receptor α1β1 suggesting that interaction of one receptor with its matrix ligand triggers a mechanism leading to stabilization of other integrin molecules at the cell surface (Dalton et al 1995).

The different results on the regulation of integrin expression by cell adhesion have not been reconciled, although it is probable that they derive from the different cellular systems and experimental conditions used.

REFERENCES

Adams JC, Watt FM (1989) Fibronectin inhibits the terminal differentiation of human keratinocytes. Nature 340:307-309.

Arner EC, Tortorella MD (1995) Signal transduction through chondrocyte integrin receptors induces matrix metalloproteinase synthesis and synergizes with interleukin-1. Arthritis Rheum 38:1304-1314.

Bissell DM, Caron JM, Babiss LE, Friedman JM (1990) Transcriptional regulation of the albumin gene in cultured rat hepatocytes. Role of basement-membrane matrix. Mol Biol Med 7:187-197.

Boudreau N, Sympson CJ, Werb Z, Bissell MJ (1995) Suppression of ICE and apoptosis in mammary epithelial cells by extracellular matrix. Science 267:891-893.

Caron JM (1990) Induction of albumin gene transcription in hepatocytes by extracellular matrix proteins. Mol Cell Biol 10:1239-1243.

Chan BM, Matsuura N, Takada Y, Zetter BR, Hemler ME (1991) In vitro and in vivo consequences of VLA-2 expression on rhabdomyosarcoma cells. Science 251:1600-1602.

Chen D, Magnuson V, Hill S, Arnaud C, Steffensen B, Klebe RJ (1992) Regulation of integrin gene expression by substrate adherence. J Biol Chem 267:23502-23506.

Coso O, Chiariello M, Yu JC, Teramoto H, Crespo P, Hu N, Miki T, Gutkind JS (1995) The small GTP-binding proteins rac1 and cdc42 regulate the activity of the jnk/sapk signaling pathway. Cell 81:1137-1146.

Dalton SL, Scharf E, Briesewitz R, Marcantonio EE, Assoian RK (1995) Cell adhesion to extracellular matrix regulates the life cycle of integrins. Mol Biol Cell 6:1781-1791.

Dhawan J, Farmer SR (1990) Regulation of alpha 1 (I)-collagen gene expression in response to cell adhesion in Swiss 3T3 fibroblasts. J Biol Chem 265:9015-9021.

Dhawan J, Lichtler AC, Rowe DW, Farmer SR (1991) Cell adhesion regulates pro-alpha 1(I) collagen mRNA stability and transcription in mouse fibroblasts. J Biol Chem 266:8470-8475.

DiPersio CM, Jackson DA, Zaret KS (1991) The extracellular matrix coordinately modulates liver transcription factors and hepatocyte morphology. Mol Cell Biol 11:4405-4414.

Eckes B, Mauch C, Huppe G, Krieg T (1993) Downregulation of collagen synthesis in fibroblasts within three-dimensional collagen lattices involves transcriptional and posttranscriptional mechanisms. FEBS Lett 318:129-133.

Eierman DF, Johnson CE, Haskill JS (1989) Human monocyte inflammatory mediator gene expression is selectively regulated by adherence substrates. J Immunol 142:1970-1976.

Fan ST, Mackman N, Cui MZ, Edgington TS (1995) Integrin regulation of an inflammatory effector gene. Direct induction of the tissue factor promoter by engagement of beta 1 or alpha 4 integrin chains. J Immunol 154:3266-3274.

Folkman J, Moscona A (1978) Role of cell shape in growth control Nature 273:345-349.

Grinnell F (1994) Fibroblasts myofibroblasts and wound contraction. J Cell Biol 124:401-404.

Hansen LK, Mooney DJ, Vacanti JP, Ingber DE (1994) Integrin binding and cell spreading on extracellular matrix act at different points in the cell cycle to promote hepatocyte growth. Mol Biol Cell 5:967-975.

Haskill S, Johnson C, Eierman D, Becker S, Warren K (1988) Adherence induces selective mRNA expression of monocyte mediators and proto-oncogenes. J Immunol 140:1690-1694.

Haskill S, Beg AA, Tompkins SM, Morris JS, Yurochko AD, Sampson-Johannes A, Mondal K, Ralph P, Baldwin AS Jr (1991) Characterization of an immediate-early gene induced in adherent monocytes that encodes I kappa B-like activity. Cell 65:1281-1289.

Hershkowiz R, Gilat D, Miron S, Mekori YA, Aderka D, Wallach D, Vlodavsky I, Cohen IR, Lider O (1993) Extracellular matrix induces tumor necrosis factor-α secretion by an interaction between resting rat CD4+ T cells and macrophages. Immunol 78:50-57.

Hill CS, Treisman R (1995) Transcriptional regulation by extracellular signals: mechanisms and specificity. Cell 80:199-211.

Hill SC, Wyne J, Treisman R (1995) The rho family of GTPase rhoA, rac1 and cdc42Hs regulate transcriptional activation by SRF. Cell 81:1159-1170.

Huhtala P, Humphries MJ, McCarthy JB, Tremble PM, Werb Z, Damsky CH (1995) Cooperative signaling by α5β1 and α4β1 integrins regulates metalloproteinase gene expression in fibroblasts adhering to fibronectin. J Cell Biol 129:867-879.

Ignotz RA, Endo T, Massague J (1987) Regulation of fibronectin and type I collagen mRNA levels by transforming growth factor-beta. J Biol Chem 262:6443-6446.

Ignotz RA, Massague J (1987) Cell adhesion protein receptors as targets for transforming growth factor-beta action. Cell 51:189-197.

Inaguma Y, Kusakabe M, Mackie EJ, Pearson CA, Chiquet-Ehrismann R, Sakakura T (1988) Epithelial induction of stromal tenascin in the mouse mammary gland: from embryogenesis to carcinogenesis. Dev Biol 128:245-255.

Juliano RL, Haskill S (1993) Signal transduction from the extracellular matrix. J Cell Biol 120:577-585.

Klein CE, Dressel D, Steinmayer T, Mauch C, Eckes B, Krieg T, Bankert RB, Weber L (1991a) Integrin alpha 2beta 1 is upregulated in fibroblasts and highly aggressive melanoma cells in three-dimensional collagen lattices and mediates the reorganization of collagen I fibrils. J Cell Biol 115:1427-36.

Klein CE, Steinmayer T, Kaufmann D, Weber L, Brocker EB (1991b) Identification of a melanoma progression antigen as integrin VLA-2. J Invest Dermatol 96(2):281-284.

Lalli E, Sassone-Corsi P (1994) Signal transduction and gene regulation: the nuclear response to cAMP. J Biol Chem 269:17359-17362.

Langholz O, Rockel D, Mauch C, Kozlowska E, Bank I, Krieg T, Eckes B (1995) Collagen and collagenase gene expression in three-dimensional collagen lattices are differentially regulated by alpha 1 beta 1 and alpha 2 beta 1 integrins. J Cell Biol 131:1903-1915.

Lee EP, Lee W, Kaetzel CS, Parry G, Bissell MJ (1985) Interaction of mouse mammary epithelial cells with collagen substrata: regulation of casein gene expression and secretion. Proc Natl Acad Sci USA 82:1419-1423.

Li ML, Aggeler J, Farson DA, Haiter C, Hassell J, Bissell MJ (1987) Influence of a reconstituted basement membrane and its components on casein gene expression and secretion in mouse mammary epithelial cells. Proc Natl Acad Sci USA 84:136-140.

Lin TH, Rosales C, Mondal K, Bolen JB, Haskill S, Juliano RL (1995) Integrin-mediated tyrosine phosphorylation and cytokine message induction in monocytic cells. J Biol Chem 270:16189-16197.

Lin TH, Yurochko A, Kornberg L, Morris J, Walker JJ, Haskill S, Juliano RL (1994) The role of protein tyrosine phosphorylation in integrin-mediated gene induction in monocytes. J Cell Biol 126:1585-1593.

Liu JK, DiPersio CM, Zaret KS (1991) Extracellular signals that regulate liver transcription factors during hepatic differentiation in vitro. Mol Cell Biol 11:773-784.

Lofquist AK, Mondal K, Morris JS, Haskill JS (1995) Transcription-independent turnover of I kappa B alpha during monocyte adherence: implications for a translational component regulating I kappa B alpha/MAD-3 mRNA levels. Mol Cell Biol 15:1737-1746.

Mackie EJ, Chiquet-Ehrismann R, Pearson CA, Inaguma Y, Taya K, Kawarada Y, Sakakura T (1987) Tenascin is a stromal marker for epithelial malignancy in the mammary gland. Proc Natl Acad Sci USA 84:4621-4625.

Mauch C, Adelmann-Grill CB, Hatamochi A, Krieg T (1989) Collagenase gene expression in fibroblasts is regulated by a three dimensional contact with collagen. FEBS Lett 250:310-305.

Mauch C, Hatamochi A, Scharffetter K, Krieg T (1988) Regulation of collagen synthesis in fibroblasts within a three-dimensional collagen gel. Exp Cell Res 178:1508-1515.

Menko AS, Boettiger D (1987) Occupation of the extracellular matrix receptor, integrin, is a central point for myogenic differentiation. Cell 51:51-57.

Minden A, Lin A, Claret FX, Abo A, Karin M (1995) Selective activation of the jnk signaling cascade and c-jun transcriptional activity by the small GTPases rac and cdc42Hs. Cell 81:1147-1157.

Miyake S, Yagita H, Maruyama T, Hashimoto H, Miyasaka N, Okamura K (1993) β1 integrin-mediated interaction with extracellular matrix proteins regulates cytokine gene expression in synovial fluid cells of rheumatoid arthritis patients. J Exp Med 177:863-868.

Pacifici R, Basilico C, Roman J, Zutter MM, Santori SA, McCracken R (1992) Collagen-induced release of interleukin 1 from human blood mononuclear cells. Potentiation by fibronectin binding to the α5β1 integrin. J Clin Invest 89:61-67.

Qwarnstrom EE, Ostberg CO, Turk GL, Richardson CA, Bomsztyk K (1994) Fibro-nectin attachment activates the NF-κB p50/65 heterodimer in fibroblasts and smooth muscle cells. J Biol Chem 269:30765-30768.

Rana B, Mischoulon D, Xie Y, Bucher NL, Farmer SR (1994) Cell-extracellular matrix interactions can regulate the switch between growth and differentiation in rat hepatocytes: reciprocal expression of C/EBP alpha and immediate-early growth response transcription factors. Mol Cell Biol 14:5858-5869.

Riikonen T, Westermarck J, Koivisto L, Broberg A, Kahari VM, Heino J (1995) Integrin alpha 2 beta 1 is a positive regulator of collagenase (MMP-1) and collagen alpha 1(I) gene expression. J Biol Chem 270:13548-52.

Romanic AM, Madri JA (1994) The induction of the 72 kDa gelatinase in T cells upon adhesion to endothelial cells is VCAM-1 dependent. J Cell Biol 125:1165-1178.

Roskelley CD, Desprez PY, Bissell MJ (1994) Extracellular matrix-dependent tissue-specific gene expression in mammary epithelial cells requires both physical and biochemical signal transduction. Proc Natl Acad Sci USA 91:12378-12382.

Saarialho Kere UK, Chang ES, Welgus HG, Parks WC (1992) Distinct localization of collagenase and tissue inhibitor of metalloproteinases expression in wound healing associated with ulcerative pyogenic granuloma. J Clin Invest 90:1952-1957.

Saarialho Kere UK, Kovacs SO, Pentland AP, Olerud JE, Welgus HG, Parks WC (1993) Cell-matrix interactions modulate interstitial collagenase expression by human keratinocytes actively involved in wound healing. J Clin Invest 92:2858-2866.

Santala P, Larjava H, Nissinen L, Riikonen T, Maatta A, Heino J (1994) Suppressed collagen gene expression and induction of alpha2beta1 integrin-type collagen receptor in tumorigenic derivatives of human osteogenic sarcoma (HOS) cell line. J Biol Chem 269:1276-1283.

Schindler C, Darnell JE Jr (1995) Transcriptional responses to polypeptide ligands: the Jak-Stat pathway. Annu Rev Biochem 64:621-51.

Schmidhauser C, Bissell MJ, Myers CA, Casperson GF (1990) Extracellular matrix and hormones transcriptionally regulate bovine beta-casein 5' sequences in stably transfected mouse mammary cells. Proc Natl Acad Sci USA 87:9118-9122.

Schmidhauser C, Casperson GF, Bissell MJ (1994) Transcriptional activation by viral enhancers: critical dependence on extracellular matrix-cell interactions in mammary epithelial cells. Mol Carcinog 10:66-71.

Schmidhauser C, Casperson GF, Myers CA, Sanzo KT, Bolten S, Bissell MJ (1992) A novel transcriptional enhancer is involved in the prolactin- and extracellular matrix-dependent regulation of beta-casein gene expression. Mol Biol Cell 3:699-709.

Seftor REB, Seftor EA, Ghelsen KR, Stetler-Stevenson WG, Brown PD, Ruoslahti E, Hendrix MJC (1992) Role of αVβ3 integrin in human melanoma cell invasion. Proc Natl Acad Sci USA 89:1557-1561.

Seftor REB, Seftor EA, Stetler-Stevenson WG, Hendrix MJC (1993) The 72 kDa type IV collagenase is modulated via differential expression of αVβ3 and α5β1 integrins during human melanoma cell invasion. Cancer Res 53:3411-3415.

Shaw RJ, Doherty DE, Ritter AG, Benedict SH, Clark RA (1990) Adherence-dependent increase in human monocyte PDGF(B) mRNA is associated with increases in c-fos, c-jun, and EGR2 mRNA. J Cell Biol 111:2139-2148.

Siebenlist U, Franzoso G, Brown K (1994) Structure, regulation and function of NF-κB. Ann Rev Cell Biol 10:405-455.

Snyder RW, Lenburg ME, Seebaum AT, Grabel LB (1992) Disruption of the cytoskeleton-extracellular matrix linkage promotes the accumulation of plasminogen activators in F9 derived parietal endoderm. Differentiation 50:153-162.

Sporn SA, Eierman DF, Johnson CE, Morris J, Martin G, Ladner M, Haskill S (1990) Monocyte adherence results in selective induction of novel genes sharing homology with mediators of inflammation and tissue repair. J Immunol 144:4434-4441.

Streuli CH, Bailey N, Bissell MJ (1991) Control of mammary epithelial differentiation: basement membrane induces tissue-specific gene expression in the absence of cell-cell interaction and morphological polarity. J Cell Biol 115:1383-1395.

Streuli CH, Bissell MJ (1990) Expression of extracellular matrix components is regulated by substratum. J Cell Biol 110:1405-1415.

Streuli CH, Edwards GM, Delcommenne M, Whitelaw CB, Burdon TG, Schindler C, Watson CJ (1995) Stat5 as a target for regulation by extracellular matrix. J Biol Chem 270:21639-21644.

Streuli CH, Schmidhauser C, Kobrin M, Bissell MJ, Derynck R (1993) Extracellular matrix regulates expression of the TGF-beta 1 gene. J Cell Biol 120:253-260.

Su B, Karin M (1996) Mitogen-activated protein kinase cascades and regulation of gene expression. Cur Op Immunol 8:402-411.

Sudbeck BD, Parks WC, Welgus H G, Pentland AP (1994) Collagen stimulated induction of keratinocyte collagenase is mediated via tyrosine kinase and protein kinase C activities. J Biol Chem 269:30022-30029.

Sympson CJ, Talhouk RS, Alexander CM, Chin JR, Clift SM, Bissell MJ, Werb Z (1994) Targeted expression of stromelysin-1 in mammary gland provides evidence for a role of proteinases in branching morphogenesis and the requirement for an intact basement membrane for tissue-specific gene expression. J Cell Biol 125:681-693.

Talhouk RS, Bissell MJ, Werb Z (1992) Coordinated expression of extracellular matrix-degrading proteinases and their inhibitors regulates mammary epithelial function during involution. J Cell Biol 118:1271-1282.

Thieblemont N, Haeffner-Cavaillon N, Haeffner A, Cholley B, Weiss L, Kazatchkine MD (1995) Triggering of complement receptors CR1 (CD35) and CR3 (CD11b/CD18) induces nuclear translocation of NF-κB (p50/65) in human monocytes and enhances viral replication in HIV-infected monocytic cells. J Immunol 155:4861-4867.

Tremble P, Chiquet-Ehrismann R, Werb Z (1994) The extracellular matrix ligands fibronectin and tenascin collaborate in regulating collagenase gene expression in fibroblasts. Mol Biol Cell 5:439-453.

Tremble P, Damsky CH, Werb Z (1995) Component of the nuclear signaling cascade that regulate collagenase gene expression in response to integrin-derived signals. J Cell Biol 129:17071720.

Unemori EN, Werb Z (1986) Reorganization of polymerized actin: a possible trigger for induction of procollagenase in fibroblasts cultured in and on collagen gels. J Cell Biol 103:1021-1031.

Vojtek AB, Cooper JA (1995) Rho family members: activator of MAP kinase cascades. Cell 82:527-529.

Watt FM, Kubler MD, Hotchin NA, Nicholson LJ, Adams JC (1993) Regulation of keratinocyte terminal differentiation by integrin-extracellular matrix interactions. J Cell Sci 106:175-82.

Werb Z, Tremble PM, Behrendtsen O, Crowley E, Damsky CH (1989) Signal transduction through the fibronectin receptor induces collagenase and stromelysin gene expression. J Cell Biol 109:877-889.

West CM, Lanza R, Rosenbloom J, Lowe M, Holtzer H, Avdalovic N (1979) Fibronectin alters the phenotypic properties of cultured chick embryo chondroblasts. Cell 17:491-501.

Xie DL, Meyers R, Homandberg GA (1993) Release of elastase from monocytes adherent to a fibronectin-gelatin surface. Blood 81:186-192.

Yamada A, Nikaido T, Nojima Y, Schlossman SF, Morimoto C (1991) Activation of human CD4 T lymphocytes. Interaction of fibronectin with VLA-5 receptor on CD4 cells induces the AP-1 transcription factor. J Immunol 146:53-56.

Yurochko AD, Liu DY, Eierman D, Haskill S (1992) Integrins as a primary signal transduction molecule regulating monocyte immediate-early gene induction. Proc Natl Acad Sci USA 89:9034-9038.

CHAPTER 8

Regulation of Cell Growth and Apoptosis

PATHWAYS REGULATING CELL CYCLE PROGRESSION

Activation of Ras-MAPK (see chapter 5) is a major pathway mediating the mitogenic response of growth factors. Activation of MAPK is required to induce the expression of "immediate early genes" such as c-jun, c-fos and c-myc. These are DNA-binding proteins required to trigger gene expression leading to synthesis of cyclins.

Cyclins play a crucial role in cell cycle progression as they bind to and regulate the activity of a class of serine/threonine kinases known as Cdk (Cyclin dependent kinases). Several different cyclins (termed A-H) and Cdks (numbered from 1 to 6) have been identified that control different phases of the cell cycle. The most relevant for our discussion are cyclins D, E and A that, in association with Cdk4/6 and Cdk2, regulate G0-G1 transition and progression along G1 (Fig. 8.1) (Lees, 1995; Morgan, 1995). The timing of expression of various cyclins is crucial in determining at which phase of the cell cycle the associated Cdk become active. Thus, cyclin abundance is rate limiting for progression through the different phases of the cell cycle. Cyclin D is synthesized in early G1 and activates Cdk4 and 6. Cyclin E-Cdk2 and cyclin A-Cdk2 form later in G1 as cells prepare to synthesize DNA. Cyclins are short lived molecules whose levels are regulated both transcriptionally and by proteolysis via the ubiquitination pathway. Cdk activity can also be regulated by phosphorylation events and by two classes of inhibitor molecules. Phosphorylation in Tyr by the Wee1 PTK has an inhibitory effect and is counteracted by the CDC25 phosphatase. Phosphorylation in Thr residues is controlled by the CAK Ser/Thr kinase and is required for activity (Fig. 8.1). Cdk activity is also regulated by inhibitory subunits. p15, p16, p18 and p19 proteins, known as the INK4 family, inhibit cyclin D-Cdk4/6 by preventing the formation of the cyclin-Cdk complex. A different group of inhibitors is represented by the p21, p27 and p57 molecules that bind to the cyclin E/A-Cdk2 complexes and regulate their activity. The stochiometry of binding is crucial as multiple copies of these molecules are required to inhibit Cdk activity. Cdk activity is, thus, controlled by the combinatorial effect of positive and negative regulatory subunits (cyclins and inhibitors) and by phosphorylation events (Fig. 8.1).

Once Cdk are activated, they trigger a cascade of phosphorylation events that ultimately releases the inhibitory machinery that refrains cells from proliferating. A major component

Signal Transduction by Integrins, by Paola Defilippi, Angela Gismondi, Angela Santoni and Guido Tarone. © 1997 Landes Bioscience.

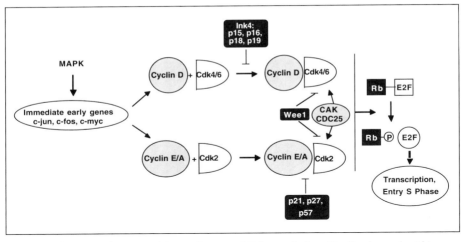

Fig. 8.1. Schematic diagram of the pathways of Cdk regulation. Cyclin dependent kinases Cdk2, Cdk4 and Cdk6 are the major kinases controlling progression along the G1 phase of the cell cycle. Cdks are regulated by both positive (light gray boxes) and negative (black boxes) signals. Cyclins function as positive regulatory subunits and activate Cdks by direct binding. Their level rises by increased gene transcription in response to mitogenic stimuli. The formation of the cyclin D-Cdk4/6 complex can be inhibited by a group of proteins named INK4, while the activity of the cyclin E-Cdk2 and cyclin A-Cdk2 complexes can be inhibited by p21, p27 and p57. All these inhibitors function by direct association with either the cyclin or the Cdk molecules. A third mechanism of regulation is represented by the phosphorylation. While tyrosine phosphorylation by Wee1 represents a negative signal, phosphorylation of a threonine residue by the action of the CAK kinase is a positive signal. Phosphorylation of a second threonine residue, however, negatively regulates the Cdk activity. Entry into the S phase of the cell cycle requires the phosphorylation of the Rb protein, a major substrate of the active cyclin-Cdk complexes. Phosphorylation of Rb allows release of the transcription factor E2F that promote gene transcription and entry into the S phase.

of this inhibitory machinery is the retinoblastoma protein Rb, a nuclear protein whose activity depends on its state of phosphorylation. De-phosphorylated Rb binds and sequesters transcription factors, such as E2F, that favor cell proliferation, while phosphorylation of Rb causes release of these proteins allowing them to promote cell cycle progression (Fig. 8.1).

INTEGRINS CAN NEGATIVELY REGULATE CELL GROWTH

The interest in the control of cell growth by extracellular matrix and integrins originates from the observation that growth of normal cells in vitro requires adhesion to the substratum, but tumor cells have lost this control (Stoker 1968).

One of the first experimental evidence pointing to a role for integrins in this event came from the analysis of tumorigenic cell lines expressing different levels of the α5β1 fibronectin receptor. Overexpression of this integrin partially reverses the tumorigenic phenotype of the cells by blocking growth in soft agar, a property typical of tumor cells (Fig. 8.2) and, most importantly, by suppressing tumor growth in nude mice (Gaincotti and Ruoslahti, 1990). The relationship between high levels of α5β1 expression and reduced in vivo growth was further confirmed by the observation that a fibronectin receptor-deficient CHO cell variant exhibits increased tumorigenicity and this property can be reversed by restoring high receptor expression levels by trans-

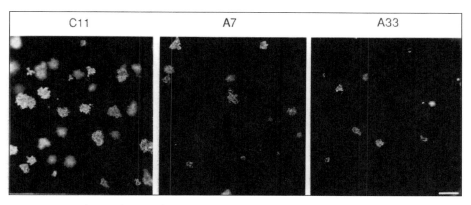

Fig. 8.2. Growth in soft agar of CHO clones expressing the human α5β1 integrin at different levels. Cells of the clone C11 (control untransfected), A7 (α5β1 intermediate expressor), and A33 (α5β1 high expressor) were plated in soft agar, and photographs of representative fields were taken after 5 days of culture. Scale bar: 100 μm. With permission from Giancotti FG and Ruoslathi E Cell 1990; 60:849-859. Copyright Cell Press.

fection (Schreiner et al 1991). The ability of α5β1 to control tumor cell growth is not restricted to CHO cells. In fact, a K562 erythroleukemia cell variant with high level of α5β1 expression proliferates slowly in liquid culture, shows a greatly decreased cloning efficiency in soft agar as compared with the parental K562, and is nontumorigenic in nude mice (Symington 1990). Moreover, tumorigenicity of α5β1-deficient human HT29 colon carcinoma cells can be suppressed by transfection of the α5β1 fibronectin receptor (Varner et al 1995).

In addition to α5β1, the α6β4 integrin was reported to inhibit growth when expressed in an α6β4-deficient rectal carcinoma cell line RKO (Clarke et al 1995).

These results suggest that the α5β1 fibronectin receptor, as well as other integrins such as α6β4, can inhibit tumor cell growth by restoring normal growth control properties. It is thus possible that these integrin receptors exert a negative control on cell growth. This is in apparent contradiction with evidence showing that occupation of α5β1 with Arg-Gly-Asp containing peptides or adhesion of cells to fibronectin or laminin triggers a positive growth signal (Symington, 1992; Mortarini et al 1992; Mainiero et al 1997). This question was addressed by Varner et al (1995) using the

α5β1-deficient HT29 colon carcinoma cells transfected with the α5β1 fibronectin receptor. In this system, it was shown that the unligated form of the α5β1 generates negative signals that suppress cell growth (Fig. 8.3), such as induction of the growth arrest specific gene-1 (gas-1) and suppression of c-fos, c-jun and junB mRNAs (Varner et al 1995). This may account for the reduced tumorigenicity in nude mice of α5β1-positive HT29 cells as compared with α5β1-negative wild type cells. At the same time, binding of the α5β1 to fibronectin in the presence of growth factors reverses gas-1 gene induction and allows immediate early gene transcription leading to cell growth (Varner et al 1995). These findings are consistent with the hypothesis that, in the absence of attachment to fibronectin, α5β1 integrin activates a signaling pathway leading to decreased tumor cell proliferation and that ligation of this receptor with fibronectin reverses this signal thereby contributing to cell proliferation. This is a rather provocative hypothesis in that it implies that an unoccupied receptor can generate an active signal suppressing cell growth, rather than simply behaving as a silent receptor.

In cells overexpressing α5β1, such as transfected CHO and the K562 variant (Giancotti and Ruoslahti, 1990; Symington,

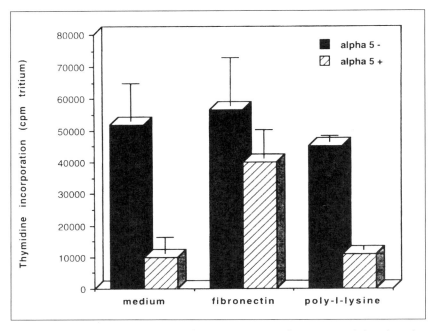

Fig. 8.3. Thymidine incorporation by HT29 α5 transfectants is inhibited in the absence of attachment to fibronectin. SVneo (black bars) and α5 positive (dashed bars) transfectants were maintained in culture in complete medium, in suspension culture in medium supplemented only with insulin, selenium and transferrin or in medium supplemented with insulin, selenium and transferrin on tissue culture plastic that had been coated with 25 µg/ml fibronectin (fibronectin) or poly-L-lysine (poly-l-lysine) in the presence of 10 µC/ml tritiated thymidine at 5 x 10⁴ cells/ml for 24 h, and acid precipitable counts were determined. Reproduced from Varner JA et al Mol Biol Cell 1995; 6:725-740, with permission of the American Society for Cell Biology.

1990), a large fraction of the receptor molecules may be in the unligated state and thus send a negative signal that partially restores normal growth properties in tumor cells.

A number of studies in different cellular systems indicate that cell matrix interaction can also negatively regulate growth and favor cell differentiation. In hepatocytes, adhesion on collagen type I, allowed growth factor-driven induction of c-jun, junB, c-fos, c-myc mRNAs and entry in S phase. However, an opposite effect was obtained with the basal membrane-like matrix produced by the Engelbreth-Holm-Swarm (EHS) mouse tumor (Rana et al 1994). This extracellular matrix allowed induction of liver-specific transcription factors and caused arrest in G_0 with inhibition of c-jun, junB

and c-myc mRNA expression. Endothelial cells proliferate when plated on fibronectin but stop growing and form capillary-like structures when plated on laminin-rich matrices (Kubota et al 1988). Similarly, myoblasts proliferate on fibronectin but fuse to form myotubes on laminin; this property has been related to the expression of specific receptors for the two matrix ligands (von der Mark and Ocalan, 1989; Sastry et al 1996). In keratinocytes, fibronectin prevents terminal differentiation allowing cell growth, while collagen IV and laminin have the opposite effect (Adams and Watt, 1989; Watt et al 1993). The basic mechanisms of these differential responses are not known at present. It has been proposed that integrins, such as α2β1 and α6β1, that are not able to activate the Ras-MAPK pathway via the adaptor protein Shc, promote exit from the cell cycle and, in the presence of appropriate stimuli, may allow entry of the differentiation pathway (Wary et al 1996) (see also chapter 5).

The ability of an integrin receptor to generate negative regulatory signals on cell proliferation has also been reported in hemopoietic cells. Fibronectin, via the α4β1 integrin receptor, may exert a negative effect on the proliferation of hemopoietic progenitor cells (Hurley et al 1995). In lymphocytes, β1 integrin was shown to suppress CD2- or CD3-mediated T cell proliferation when stimulated with the K20 monoclonal antibody (Tichioni et al 1993; Groux et al 1989). This effect is accompanied by increase in cAMP (Groux et al 1989) and decrease in diacylglycerol and phosphatidic acid levels (Tichioni et al 1993). Similar results were also reported using another β1 monoclonal antibody 18D3 (Bednarczyk et al. 1992; Teague and McIntyre, 1994). Several other antibodies to β1 integrin, however, have no effect or actually stimulate CD3-induced T cell proliferation as discussed below (Martsuyama et al 1989; Dang et al 1990; Nojima et al 1990; Shimizu et al 1990; Yamada et al 1991). Thus, it is likely that specific β1 ligands can trigger different effects on cell proliferation.

SPECIFIC INTEGRIN SUBUNITS CAN REGULATE CELL GROWTH

Although the molecualr mechanisms of negative regulation of cell growth by integrins are still poorly understood, a number of studies indicates that specific b subunit cytoplasmic domain sequences can play distinct roles in the control of cell cycle progression.

β1 and β5 cytoplasmic domains apparently deliver different signals in response to antibody binding. The β1 molecule generates a proliferative response when stimulated by a combination of fibronectin and β1 antibody, whereas the chimeric molecule with the extracellular portion of β1 and the cytoplasmic sequence of β5 does not (Pasqualini and Hemler, 1994).

β1C, an alternative splice variant of the β1 subunit characterized by a specific cytoplasmic domain (see chapter 1), is also able to suppress cell growth. Microinjection of β1C cDNA in quiescent C3H10T1/2 cells inhibited DNA synthesis upon serum stimulation (Meredith et al 1995). β1C seems to act at late G1 by preventing entry in S phase since its expression blocks the appearance of cyclin A, but does not affect appearance of cyclin E (Meredith et al 1995). Contradictory results have been reported concerning the active amino acid sequence responsible for the negative regulation of cell growth. In fact, two different sequences were identified. In one case, the growth arrest capacity resides in the 13 most COOH terminal residues (Meredith et al 1995), while in another report, a more upstream region encompassing residues Gln-795 and Gln-802 was found to be functionally important (Fornaro et al 1995). Interestingly, expression of β1C correlates with a growth arrested phenotype in cytokine-stimulated vascular endothelial cells suggesting that the specific functional property of β1C are physiologically relevant (Fornaro et al 1995).

Other integrins, such as the β6 subunit, have a stimulatory effect on cell growth. The αVβ6 integrin is a fibronectin receptor rarely detected in adult tissues, but is expressed during fetal development, wound healing and in a variety of epithelial tumors (Agrez et al 1994). This pattern of expression suggests a possible role of αVβ6 integrin in epithelial cell proliferation. Indeed expression of β6 in colon carcinoma cell line SW480 that normally lacks this integrin, enhances the proliferative capacity of these cells in tri-dimensional collagen gels and in vivo in nude mice (Fig. 8.4). Mutational analysis showed that a region of 11 amino acids at the COOH terminus of the β6 cytoplasmic domain is required for this proliferative effect. This region is unique to the β6 cytoplasmic domain and similar sequences are not present in the corresponding domain of β1, β2 and β3 subunits that are otherwise highly homologous. Removal of this 11 amino acid sequence does not affect the adhesive property of β6 nor its localization to

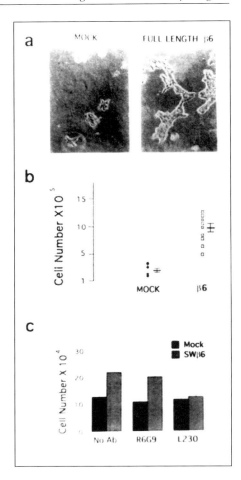

Fig. 8.4. Effect of β6 transfection on the ability of SW480 cells to proliferate within three-dimensional collagen type I gel. (a) Photomicrograph of mock and full length β6 transfectants after 7 days in collagen gel. (b) Total cell counts for each of five mock clones and nine full-length β6 clones. Cells were removed from the gel by treatment with collagenase and counted in a hemocytometer. Bars to the right of each set of individual data points describe the mean SD. (c) Effect of blocking antibodies against the αV (L230) on total cell counts of mock and β6 transfected cells allowed to proliferate for 7 days in collagen gels. The nonblocking anti-β6 antibody, R6G9, was used as control. Shaded bars represent mean values of triplicate wells and lines above each bar represent the standard error of the mean. From Agrez M et al J Cell Biol 1994; 127:547-556 by copyright permission of The Rockefeller University Press.

focal contacts, but it abrogates the ability to stimulate proliferation in vivo and in tridimensional collagen gels. Thus, this β6-specific extra-sequence is not related to adhesive function, but rather it controls specific responses related to cell growth.

The data discussed above, thus, indicate that integrin cytoplasmic sequences generate signals that either positively or negatively control cell proliferation, suggesting that regulation of integrin expression can be used by cells to finely regulate their proliferative response to extracellular stimuli.

INTEGRINS PROMOTE CELL GROWTH BY REGULATING CYCLINS AND Rb PROTEIN

Several studies indicate that adhesion to matrix proteins can favor cell growth in vitro

(Ingber 1990; Panayotou et al 1989; Mortarini et al 1992 and 1995). This is compatible with a positive regulatory role of integrins in this process. Upon cell-matrix adhesion in the absence of mitogenic stimuli from growth factors, β1, αV and β4 integrins are able to trigger the Ras-MAPK pathway likely providing a positive signal for cell growth (Chen et al 1994; Schlaepfer et al 1994; Zhu and Assoian, 1995; Renshaw et al 1996; Wary et al 1996; Mainiero et al 1997) (see chapter 5). Additionally, growth factors, such as EGF, PDGF and bFGF, can also trigger the Ras-MAPK pathway in cells that are not in contact with the matrix, thus confirming that both integrins and growth factors can independently activate this pathway (Zhu and Assoian, 1995; Myamoto et al 1996).

A number of questions arise. Why are two independent pathways necessary to control Ras-MAPK activation in cell growth? Are the integrin and growth factor signaling diverging downstream of the MAPK? Are both required for a growth response?

A recent report showed that the action of integrins and growth factors, such as EGF, PDGF-BB and bFGF, on Erk activation is additive (Myamoto et al 1996), suggesting that the combined signaling of integrin and growth factors may be necessary for progression of the signaling cascade.

Analysis of the downstream signaling, moreover, identified some differences in the two pathways. Integrin-dependent activation of MAPK Erk1 and Erk2 leads to transcription of c-fos gene (Wary et al 1996; see also chapter 7) consistent with previous findings showing matrix-induced transcription of the immediate early genes c-fos and c-myc in serum starved 3T3 cells (Dike and Farmer, 1988). However, bFGF can not efficiently trigger this response in nonadherent endothelial cells suggesting that the two pathways diverge and integrins, but not growth factors, can induce c-fos gene transcription (Wary et al 1996). The two stimuli can synergize and the costimulation of integrins and bFGF receptor strongly stimulates c-fos gene transcription with respect to the integrin stimulus alone (Wary et al 1996) (Fig. 8.5). Analysis of the downstream steps of the pathway showed that synthesis of cyclin D and A and activation of the cyclin E/Cdk2 complex require the simultaneous presence of growth factor and adhesion signals for cell cycle progression (Fig. 8.6) (Zhu et al 1996; Fang et al 1996). Adhesion-dependent regulation of the three cyclins occurs by different mechanisms. While cyclins D and A appear to be regulated at the biosynthetic level (Zhu et al 1996), cyclin E is regulated at posttransductional level (Zhu et al 1996; Fang et al 1996). Cyclin E and the Cdk2 levels are not affected by adhesion, but the cyclin E/Cdk2 kinase is inactive in suspended, growth factor-stimulated cells (Zhu

et al 1996; Fang et al 1996). This is due to tight association of the cyclin E/Cdk2 complex with two inhibitor proteins, p21 and p27. These inhibitors are released and cyclin E/Cdk2 kinase activity can be measured only if cells are allowed to adhere. Cyclins D and A, on the other hand, are regulated at the mRNA level since the amount of their transcripts increases progressively in adherent cells stimulated by mitogens. Induction of cyclin D and A expression by cell adhesion is instrumental to progression through the cell cycle. In fact, enforced expression of either cyclins in late G1 induces entry in S phase independently from cell-substratum adhesion (Guadagno et al 1993; Zhu et al 1996).

A second event important for G1 progression and entry in S phase is the phosphorylation of the Rb protein. This event is also adhesion-dependent (Symington, 1992; Zhu et al 1996) which is consistent with the hypothesis that Rb is a substrate of Cdk activity (Fig. 8.6).

Taken together these results suggest that the combined action of adhesion and growth factors regulates cyclin D and cyclin A expression leading to the formation of the active cyclin D/Cdk4 and cyclin A/Cdk2 complexes. At the same time, activation of the cyclin E/Cdk2 kinase occurs by dissociation of the inhibitory p21 and p27 proteins. These events can lead to Rb protein phosphorylation that is likely to induce entry in S phase.

Adhesion-dependent progression along G1-S seems to require organization of the actin cytoskeleton and cell spreading. Disruption of the actin cytoskeleton with cytochalasin D in G0 arrested fibroblasts prevents progression along G1 and entry into S phase (Bohmer et al 1996). Organization of the cytoskeleton and cell spreading is also required for induction of junB and ras genes and G1-S transition in primary hepatocytes (Hansen et al 1994).

The data discussed above offer a molecular basis for a phenomenon commonly observed during in vitro cell culture of epithelial and most mesenchymal cells, known as anchorage-dependent cell growth (Stoker, 1968). Normal cells kept from adhering to the growth surface by suspension in soft agar or methylcellulose, are arrested in most macro-molecular biosynthetic processes, such as RNA and protein synthesis, even in the presence of growth factors and metabolites (Benecke et al 1978; Farmer et al 1978; Ben-Ze'ev et al 1980). Moreover, these cells do not synthesize DNA and are, thus, growth arrested (Otsuka and Moskowitz, 1975; Matsuhisa and Mori, 1981). Using culture dishes coated with increasing concentrations of hydrophobic, nonadhesive polymer, polyhydroxyethyl-methacrylate (poly-HEMA), it was possible to generate substrata that allow different degrees of cell adhesion and spreading. In this system, the growth rate, as measured by DNA synthesis, is directly proportional to the degree of cell spreading on the culture substratum (Folkman and Moscona, 1978). These experiments clearly established a strict relationship between adhesion to a substratum and ability of a cell to grow in response to growth factors. The recent findings discussed above provide insight at molecular level explaining why growth factors cannot induce cell growth unless accompanied by integrin signaling. The two signaling pathways, by converging on MAPK, may induce a threshold activity necessary to elicit significant c-fos transcription and cyclin induction (Fig. 8.7).

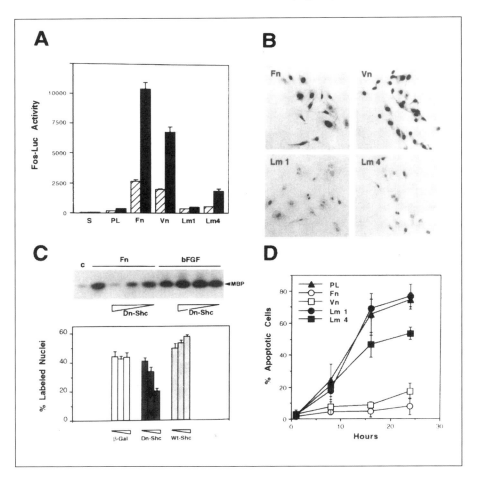

INTEGRIN-DEPENDENT REGULATION OF CYCLINS IS ABNORMAL IN NEOPLASTIC CELLS

Neoplastic cells are characterized by the loss of anchorage-dependent cell proliferation. Tumor cells, in fact, grow in suspension, a property which is the best in vitro correlate of tumorigenicity (Shin et al 1975). Analysis of the growth requirements of the anchorage-independent, chemically transformed HUT12 cell line indicated that these cells have lost the adhesion-dependent induction of cyclin E/Cdk2 activity (Fang et al 1996). Moreover, NRK cells, which are immortalized but not tumorigenic, have lost the adhesion requirement for the expression of cyclin D and induction of cyclin E/Cdk2 activity, but not for the expression of cyclin A (Zhu et al 1996). Thus, the loss of anchorage dependence for growth in neoplastic cells can arise as a multi-step deregulation of the adhesion-controlled cell cycle events. A possible deregulation mechanism has been proposed in v-Src-transformed cells. In these cells, p125Fak is constitutively phosphorylated by v-Src and bound to Grb2/Sos complex (Schlaepfer et al 1994). It is possible that this complex controls v-Src-mediated MAPK activation bypassing the integrin pathway and rendering cell growth independent from adhesion.

Fig. 8.5 (opposite) Control of Fos-SRE-dependent transcription, cell survival and cell cycle progression by the extracellular matrix. (A) Human umbilical vein endothelial cells (HUVECs) were transiently transfected with fos-SRE-luciferase plasmid. After growth factor starvation, the cells were either kept in suspension (S) or plated onto dishes coated with 10 µg/ml poly-L-lysine (PL), fibronectin (Fn), laminin 1 (Lm1) or laminin 4 (Lm4) for 20 min. The cells were then either left untreated (hatched bars) or exposed to 10 ng/ml bFGF and 1 µg/ml heparin (closed bars) for 5 min. Cell lysates containing equal amounts of total proteins were subjected to luciferase assay. Values are expressed in arbitrary units. Note that bFGF in absence of matrix proteins (closed bar on PL) does not stimulate fos-SRE transcription. Transcription, however, is induced either by fibronectin alone (hatched bar, Fn) or in combination with bFGF (closed bar, Fn).
(B) HUVECs were growth factor-starved and plated for 24 hr in defined medium containing 10 µM BrdU onto wells coated with 10 µg/ml fibronectin (Fn), vitronectin (Vn), laminin 1 (Lm1) or laminin 4 (Lm4). After immunostaining with anti-BrdU monoclonal antibodies and alkaline phosphatase-conjugated secondary antibodies, the cells were lightly counterstained with hematoxylin.
(C) HUVECs were transiently transfected with 3 µg of HA-tagged-Erk2 plasmid alone or in combination with 10, 5 and 2.5 µg of plasmid encoding dominant-negative Shc (Dn-Shc). After growth factor starvation, the cells were detached and kept in suspension (c) or plated on fibronectin-coated dishes (Fn), or they were not detached, but treated with 25 ng/ml bFGF and 1 µg/ml heparin for 5 min (bFGF). Transfection efficiencies were verified by immunoblotting with anti-HA monoclonal antibodies. Immunoprecipitated HA-Erk2 was subjected to in vitro kinase assay with myelin basic protein (MBP) as substrate (top). HUVECs were transiently transfected with 10, 5 and 2.5 µg of plasmid encoding HA-tagged β-galactosidase, FLAG-tagged dominant-negative Shc (Dn-Shc), or FLAG-tagged wild-type Shc (WT-Shc). The percentage of transfected cells entering into S phase was determined and the diagram shows the mean value and standard deviation from triplicate samples (bottom).
(D) G_0-synchronized HUVECs were plated on wells coated with 10 µg/ml fibronectin (Fn), vitronectin (Vn), laminin 1 (Lm1) or laminin 4 (Lm4). Adherent cells were incubated in defined medium for the indicated times. At each time point, attached and unattached cells were combined and stained in suspension with Hoechst dye. The percentage of apoptotic nuclei was determined by scoring at least 500 cells from five microscopic fields. Diagram indicates the mean and standard deviation from triplicate samples. With permission from Wary KK et al Cell 1996; 87:733-744. Copyright Cell Press.

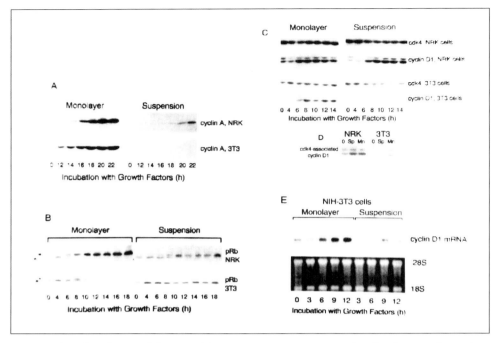

Fig. 8.6. Phosphorylation of the retinoblastoma protein is associated with adhesion-dependent expression of cyclin D1. NRK and NIH3T3 cells were synchronized in G_0, trypsinized and seeded with soluble mitogens (fetal calf serum and EGF) in monolayer or suspension. Cell were collected and extracted at the times shown. The extracts were fractionated on SDS-PAGE and analyzed by immunoblotting with antibodies to cyclin A (A), Rb (B), cyclin D1 (C) and Cdk4 (C). The upper and the lower arrowheads in B, respectively, show the hyper- and hypo-phosphorylated forms of Rb. In D, an anti-Cdk4 antibody was used to harvest cyclin D-Cdk4 complexes from extracts (0.5 mg) of monolayer (Mn) and suspension (Sp) NRK and NIH3T3 cells. The immunoprecipitates were fractionated on reducing SDS-PAGE and the amount of associated cyclin D1 was determined by western blotting. In E, equal amounts of total RNA (assessed by ethidium bromide staining of rRNA) were isolated from extracts of monolayer and suspended NIH3T3 cells and analyzed by northern blot hybridization with a murine cyclin D1 cDNA. From Zhu X et al J Cell Biol 1996; 133:391-403 by copyright permission of The Rockefeller University Press.

Tumor cells can also proliferate in response to matrix stimuli in the absence of exogenous growth factors. Melanoma cells in vitro can be induced to proliferate by plating on fibronectin (Mortarini et al 1992) or on laminin in the absence of exogenous growth factors (Mortarini et al 1995). This does not apply to untransformed melanocytes whose proliferation requires both laminin and growth factors. A possible explanation is that some tumor cells have a constitutively activated mitogen signaling pathway and, thus, triggering of integrin sig-

naling is sufficient to reach a threshold level of stimulation. These data suggest that neoplastic transformation can arise in consequence of the alteration of either one of the signaling pathways triggered by integrins or growth factor receptors schematized in Figure 8.7.

LYMPHOCYTE PROLIFERATION IN RESPONSE TO ANTIGEN IS POTENTIATED BY INTEGRINS

The request of matrix and growth signals to trigger cell proliferation has also been

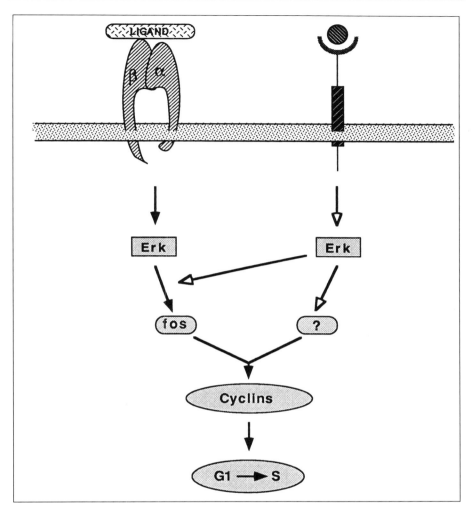

Fig. 8.7. Cooperation of integrin and growth factor signaling in the mitogenic response. Matrix proteins via their integrin receptors trigger Erk MAPK activation and fos gene transcription. Growth factors such as bFGF, although able to trigger Erk MAPK activation, activate fos gene transcription only in the presence of a costimulus from the extracellular matrix (see Wary et al 1996). This may be responsible for induction of cyclins synthesis and entry into the S phase of the cell cycle. Whether Erk activation via growth factor receptors activate some molecules that cooperate with fos to regulate cyclin expression is not known (?).

reported for lymphocytes. Both α3β1, α4β1, α5β1 and α6β1 integrin complexes and their respective matrix ligands, collagen, fibronectin and laminin, promote CD3-dependent proliferation of T cells (Martsuyama et al 1989; Dang et al 1990; Nojima Y et al 1990; Shimizu Y et al 1990; Yamada et al 1991). In this case, DNA synthesis requires

stimulation of both the CD3 T cell receptor molecule and integrins via specific ligands (Nojima et al 1990). Thus, as in adherent cells, proliferation of lymphocytes seems to require the convergence of both growth and adhesion signals. The requirement of a matrix signal to control proliferation may be regarded as a positional signal. Proliferation of lymphocytes in response to the antigenic stimuli occurs in lymphoid organs rather than in circulation and, thus, cells should sense the tissue environment to generate a correct response to antigens. Lymphocyte response to integrin occupancy, however, does not always result in antigen receptor-dependent proliferation. Depending on the state of cell activation, programmed cell death (apoptosis) can occur in lymphocytes in response to integrin stimulation (see below)

INTEGRINS MAY SYNERGIZE WITH MITOGENIC STIMULI IN SEVERAL WAYS

Recent evidences indicate that cell adhesion and integrins may affect the growth factor receptor response. It has been reported that the PDGFβ (Sundberg and Rubin, 1996) and HGF (Wang et al 1996) receptors are tyrosine phosphorylated in a ligand-independent way in response to cell-matrix adhesion or integrin clustering. The mechanism for such activation are still obscure and may differ for each receptor. It has been reported, in fact, that while spreading is required for activation of the PDGFβ receptor in human diploid foreskin fibroblasts (Sundberg and Rubin, 1996), actin cytoskeleton assembly and spreading are not required for HGF receptor tyrosine phosphorylation in B16 mouse melanoma cells (Wang et al 1996). The relevance of adhesion-dependent activation of growth factor receptor for downstream signaling is not clear at present. It may be argued that this event is important in integrin-induced MAPK activation. It has been reported, in fact, that seven membrane-spanning receptors, such as LPA, endothelin or thrombin receptors, can induce EGF receptor activation in the absence of EGF and this event is crucial for MAPK activation and for the mitogenic response to LPA (Daub et al 1996). An alternative possibility is that the adhesion-dependent growth factor receptor activation is important in the ability of integrins to generate a survival signal preventing apoptosis induced by growth factor deprivation (see below). A second important

effect of cell adhesion on growth factor receptor activation is the ability to strongly stimulate their tyrosine phosphorylation in response to the physiological ligand. It has been shown that both EGF-(Miyamoto et al 1996) and HGF-(Wang et al 1996) induced tyrosine phosphorylation of their corresponding receptors is strongly potentiated by cell adhesion, indicating that full activation of the growth factor receptors requires cell-matrix adhesion. This can be a possible mechanism explaining why both matrix and growth factors are required to reach a threshold level of MAPK activation to trigger cyclin D, E and A and cell proliferation.

The data discussed above suggest a working model in which cell-matrix adhesion via integrins leads to a basal level of growth factor receptor activation which is required for a full response to its natural growth factor ligand. In this hypothetical model growth factor receptors act as downstream effectors of integrins. Thus an alternative pathway to that proposed in Figure 8.7 predicts a potentiation of growth factor receptor signaling by integrins at the very first step of the cascade.

The MAPK pathway may not represent the only way by which integrins can synergize with mitogenic stimuli to control cell cycle progression. As discussed in previous chapters, integrins can also activate a number of other signaling events that are common to most mitogenic stimuli. As discussed in the previous chapters, these include alkalinization of the cytoplasm via the Na^+/H^+ antiport (Ingber et al 1990; Schwartz et al. 1991), activation of phosphoinositide metabolism (McNamee et al 1993), induction of Ca^{2+} transients (Schwartz, 1993) and activation of protein kinase C (Vuori and Ruoslahti, 1993) and of PI-3K (Chen and Guan, 1994).

Moreover, integrins can synergize with PDGF signaling by regulating the levels of $PtdIns(4,5)P_2$. As discussed in chapter 3, integrins can increase the intracellular levels of $PtdIns(4,5)P_2$ by stimulating the PtdIns-5-kinase via Rho (Chong et al 1994). Integrin-dependent accumulation of $PtdIns(4,5)P_2$ is required for efficient phosphatidylinositol hydrolysis (McNamee et al 1993) and cytoplasmic alkalinization (Schwartz et al 1991) upon PDGF stimulation. In this system, integrins can thus provide abundant substrate for the mitogen-triggered phospholipase C and allow a full mitogenic response to PDGF.

The $\alpha V\beta 3$ integrin was shown to associate with the insulin receptor substrate IRS-1 (Vuori and Ruoslahti, 1994). This is an adaptor protein that upon tyrosine phosphorylation can serve as a docking site for proteins, such as Grb2 and PI-3K. The $\alpha V\beta 3$-IRS-1 complex potentiates the mitogenic response of cells to insulin, thus providing another mean of integrin-dependent control of cell growth. More recently, it has been shown that $\beta 1$ integrin cytoplasmic domain interacts with a novel Ser/Thr kinase termed ILK (Integrin Linked Kinase). Endogenous ILK kinase activity is reduced in response to fibronectin cell adhesion and overexpression of ILK inhibits adhesion to matrix substrates, while inducing anchorage-independent growth (Hannigan et al 1996).

All these findings suggest that integrins may use multiple mechanisms to affect the signaling pathway triggered by mitogenic stimuli and thus control cell growth.

INTEGRINS CAN REGULATE APOPTOSIS

An additional mechanism by which integrins can affect cell growth is by regulating apoptosis. Apoptosis, a process whereby cells are induced to activate a genetic program leading to their own death, can occur as a physiological phenomenon regulating tissue homeostasis or in response to cell stress and accumulation of damaged DNA. Physiological apoptosis is regulated by the presence of signals, such as the Fas ligand or TNF-α, two cytokines that specifically induce apoptosis (Nagata and Golstein, 1995), or by the absence of stimuli, represented by either growth/differentiation factors or extracellular matrix molecules, that, when present, prevent entry into the apoptotic pathway (Raff, 1992).

In the case of stress-induced apoptosis, alterations in the growth control mechanisms or in the DNA molecule induce the expression of the p53 protein (Lowe and Ruley, 1993; Hermeking and Eich, 1994) that plays a central role in regulating growth arrest and apoptosis (Fig. 8.8). p53 can induce synthesis of p21 (El-Deiry et al 1994) that inhibits G1 cyclin/Cdk complexes (see also Fig. 8.1) leading to growth arrest. p53 can also induce the expression of Bax (Miyashita et al 1994; Miyashita and Reed, 1995), a molecule that favors apoptosis by antagonizing Bcl-2 (Reed, 1995). Bcl-2 is able to prevent apoptotic death and can act as an oncogene, most likely by inhibiting apoptotic death of tumor cells. The ratio between Bax and Bcl-2 determines the cellular response (Oltvai et al 1993).

The biochemical events downstream of the Bcl-2/Bax molecules have not been elucidated yet, but the family of ICE (*Inter-leukin Converting Enzyme*)-like proteases are involved in this pathway (Patel et al 1996), since inhibitors of these enzymes prevent apoptotic cell death (Komiyama et al 1994; Bump et al 1995) (Fig. 8.8). ICE-like proteases act on cellular proteins including nuclear lamins and enzymes involved in the DNA repair, such as PARP (Poly ADP-Ribose Polymerase) triggering cell self destruction (Patel et al 1996).

The pathway triggered by the Fas and TNF-R does not seem to involve p53 (Fig. 8.8). These membrane proteins contain in their cytoplasmic portion a protein-protein interaction motif called the "death" domain that allow binding to adaptor molecules, such as FADD (Fas Associated Death Domain protein) and TRADD (TNF-R Associated Death Domain protein). These adaptor proteins bind to a class of ICE-like proteases called MACH or FLICE leading to the assembly of a membrane-bound transduction complex (Boldin et al 1996; Muzio et al 1996). Additional events, however, may be required, since activation of these proteases and the final apoptotic effect of Fas and TNF-R can be prevented by overexpression of Bcl-2, indicating that this molecule is also involved in this pathway.

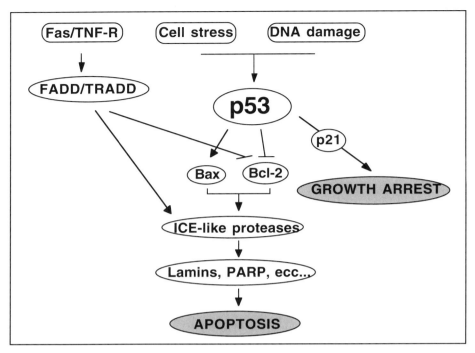

Fig. 8.8. Schematic model of the pathways leading to apoptosis. Apoptosis can be induced by the cell surface receptors, such as Fas and TNF-R, or by deprivation from growth factors or by UV and γ radiation leading to damaged DNA. Fas and TNF-R via their "death" domains interact with adaptor molecules, such as FADD and TRADD that bind ICE-like proteases. Activation of these proteases and the apoptotic effect of Fas can be counteracted by Bcl-2 indicating that this molecule is part of this pathway. The other pathway acts via p53 that in turn regulate Bax and Bcl-2 levels. p53 can regulate both cell growth and apoptosis by upregulating the p21 inhibitor of cyclin-Cdk complexes (see Fig. 8.1) and by increasing the levels of Bax, that antagonizes Bcl-2 a molecule with an anti-apoptotic function. The molecular mechanisms of Bax and Bcl-2, function are not known at present. When the Bax/Bcl-2 ration is in favor of Bax, ICE-like proteases are induced that represent the executors of the suicide program leading to apoptosis.

A role of the MAPKs in apoptosis has also been demonstrated. Activation of the Jnk and p38 and inhibition of Erk kinases (see chapter 5) upon NGF withdrawal induces apoptosis in PC12 cells (Xia et al 1995). Activation of Jnk kinases has been associated with apoptosis in several other cell types. These enzymes are likely to act upstream of p53; however, recent studies indicate that they may also be downstream of ICE-like proteases suggesting the existence of a positive feedback loop between the ICE and the Jnk system (Frisch et al 1996a, Cahill et al 1996).

Interaction of cells with the extracellular matrix protects them from apoptosis. This is true for endothelial (Meredith et al 1993; Re et al 1994), epithelial (Frisch and Hunter, 1994; Boudreau et al 1995; Saelman et al 1995) and neuronal cells (Bozzo et al 1997) (Fig. 8.9). Eosinophils and B lymphocytes can also be protected from apoptosis by interaction with fibronectin or the adhesive receptors ICAM-1 and VCAM-1 on antigen-presenting cells (Anwar et al 1993; Koopman et al 1994). Apoptosis can be induced by in vitro growth conditions that prevent attachment to the culture dish.

These include coating with saturating amounts of bovine serum albumin (Meredith et al 1993), with the synthetic polymer poly-HEMA (Frisch and Hunter, 1994; Re et al 1994) or with agarose (Meredith et al 1993). All these conditions prevent cell anchorage to the culture dish and induce various degrees of apoptosis ranging from 20% to 80% depending on the cell type and on the length of time in suspension. The first signs of apoptotic death usually occur at 8 h and maximal effects are observed at 24h.

Cell matrix adhesion may be a physiological mechanism regulating apoptosis as suggested by the mammary gland system. Epithelial cells of the mammary gland undergo involution at the end of the lactation period. At this stage production of metalloproteases that degrade the extracellular matrix is strongly increased. Expression of a metallo-protease, stromelysin, in the mammary gland in transgenic mice partially mimics these effects and induces apoptosis (Sympson et al 1994; Boudreau et al 1995). Loss of matrix contacts in these cells causes expression of ICE-like enzyme (Boudreau et al 1995), which is a key step in the apoptotic program. Apoptosis may also be an important mechanism in cell positioning during development and tissue regeneration (Raff, 1992). The fact that endothelial, epithelial and neuronal cells can survive only if bound to the appropriate extracellular matrix may be crucial in determining their positioning during development and organization in mature tissues.

Integrins are directly involved in the control of apoptosis by cell-matrix interaction. Suspended endothelial cells can be rescued from apoptosis by adherence on dishes coated with β1 integrin antibodies, but not on dishes coated with antibodies to other adhesion receptors or to histocompatibility antigens, implying that integrins are specifically involved in the generation of signals that prevent apoptosis (Meredith et al 1993). Integrin-induced tyrosine phosphorylation may indeed represent one of these signaling mechanisms. In fact, inhibition of tyrosine phosphatases by orthovanadate prevents

apoptosis in suspended cells (Meredith et al 1993). Evidence indicates that p125Fak is likely to be involved in this pathway. In fact, microinjection of p125Fak antibodies in embryo fibroblasts induces their rounding and apoptosis (Hungerford et al 1996). Attenuation of p125Fak expression induces apoptosis in tumor cells (Xu et al 1996). In chicken embryo fibroblasts, p125Fak undergoes proteolysis upon c-Myc induced apoptosis and this event is suppressed by integrin signaling (Crouch et al 1996). Moreover, expression of a constitutively activated, membrane-bound form of p125Fak consisting of the Fak sequence fused to the extracellular and transmembrane region of CD2/Fak, rescued MDCK and HaCat epithelial cell lines from apoptosis induced by loss of matrix adhesion (Frisch et al 1996b). This effect is dependent on the kinase activity of the CD2/Fak molecule, since a kinase dead CD2/Fak mutant is ineffective. In addition, mutation of the Tyr-397, the major phosphorylation site of p125Fak which serves as a binding site for SH2 domains of Src-family kinases, also reduces the anti-apoptotic activity suggesting the importance of CD2/Fak interaction with Src family kinases. Involvement of p125Fak in apoptosis, however, may depend on the cell type studied, since a mutated form of the α5β1 integrin lacking the α5 cytoplasmic domain still activates p125Fak, but no longer rescues CHO cells from detachment-induced apoptosis. Upregulation of Bcl-2 also seems to be important in integrin-mediated cell survival. Forced expression of Bcl-2 in epithelial cells can rescue apoptosis induced by loss of matrix adhesion (Frisch and Hunter, 1994). Similar effects were observed in CHO cells (Zhang et al 1995) where increased expression of Bcl-2 gene and protein is specifically induced by α5β1 fibronectin ligation. In these cells the αVβ1 integrin, which also acts as an Arg-Gly-Asp dependent fibronectin receptor, does not support cell survival (Zhang et al 1995). In neuroblastoma cells, both α1β1, α3β1 and αVβ3 integrins are able to suppress apoptosis induced by cell-substratum detachment (Bozzo et al 1997). Moreover, in human endothelial cells,

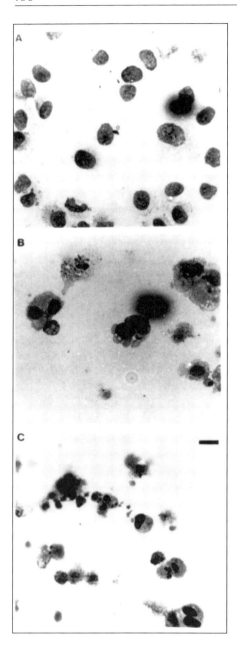

Fig. 8.9. Morphological characteristics of endothelial cells cultured under non-adherent conditions. Cells were cytocentrifuged after 12 h (B) or 18 h (C) of culture on poly HEMA coated wells, stained with May-Grunwald-Giemsa, and then examined at final magnification of 400x. Cells showed typical characteristics of apoptosis (nuclear condensation and fragmentation, surface blebbing and reduction in cell size). For comparison, an aliquot of the same cell suspension was plated in gelatin-coated dishes, cultivated for 18 h and then detached and cytocentrifuged (A). Identical results were obtained in three different experiments. Bar, 40 μm. From Re F et al J Cell Biol 1994; 127:537-546 by copyright permission of The Rockefeller University.

and this ability may depend on the cell type. αVβ3 is the most important integrin involved in this process in endothelial cells. Antibodies that specifically block ligand binding of this integrin induce apoptosis of endothelial cells in vivo and block tumor-induced angiogenesis in a model system (Brooks et al 1994). αVβ3, in addition, also protects melanoma cells from apoptosis when grown in a tridimensional collagen matrix in vitro. Since expression of this integrin correlates with acquisition of the invasive phenotype, it is possible that αVβ3 plays an important role in determining the survival of these tumor cells in the stromal environment, thus contributing to their malignancy (Montgomery et al 1994).

Clustering of integrins with secondary antibodies is sufficient to prevent apoptosis in suspended neuroblastoma cells implying that adhesion and spreading on a growth substratum are not required and integrins can per se generate a signal protecting from apoptosis (Bozzo et al 1997). This, however, may not be a general property of all cell types, since soluble integrin ligands are not sufficient to prevent apoptosis in endothelial cells, for which, spreading on a substratum is required (Re et al 1994).

The apoptotic pathway interfaces with that of growth factors and oncogenes. In fact, expression of oncogenes such as Ras and Src can counteract the apoptotic death

ligation of integrins unable to trigger the Ras-MAPK pathway via Shc, such as α6β1 causes apoptosis unless other survival stimuli are provided (Wary et al 1996). Thus, several, but not all, integrin heterodimers can generate signals controlling cell death induced by loss of cell matrix adhesion in endothelial and epithelial cells (Re et al 1994; Frisch and Hunter, 1994). Neoplastic transformation also abolishes apoptosis induced by loss of matrix contact in human breast epithelial cells (Howlett et al 1995). Since tumor cells have lost the anchorage dependence for growth it is not surprising that they do not undergo apoptosis once detached from the matrix. This may represent an important mechanism by which neoplastic cells are capable to survive in an environment different from that of the tissue of origin. Resistance to apoptosis, however, is not a general property of all malignant cell lines since neuroblastoma (Bozzo et al 1997), melanoma (Montgomery et al 1994) and CHO cells (Zhang et al 1995), known to be tumorigenic in vivo, were shown to undergo the process of self destruction upon detachment from the matrix. Interestingly, these cells can be rescued from apoptosis by growth factors present is serum, that are inactive on normal endothelial and epithelial cells (Meredith et al 1993; Re et al 1994; Frisch and Hunter, 1994; Zhang et al 1995 and Bozzo et al 1997). Thus, while in normal epithelial and endothelial cells, loss of matrix adhesion leads to obligatory triggering of the apoptotic pathway, in tumor cells this response is either abrogated or can be abrogated by exposure to growth factors.

In some cases, leukocytes can respond to integrin occupancy by undergoing apoptosis rather that proliferating or surviving. Growth of myeloid leukemia cell line M07E is suppressed by seeding on fibronectin-coated substrata due to triggering of programmed cell death (Sugahara et al 1994). This response is common to several hemopoietic cell lines and depends on the interaction of fibronectin with the $\alpha 5\beta 1$ integrin.

Costimulation of the antigen receptor CD3 with LFA-1 or $\alpha 4\beta 1$ integrins induces apoptosis in antigen-primed T helper lymphocytes (Damle et al 1993; Matsumoto et al 1994). Integrin-induced apoptosis in these cells can not be counteracted by interleukin-2 or interleukin-4 (Damle et al 1993) and is independent of the Fas path-

way (Matsumoto et al 1994). This is a striking effect since, as discussed above, costimulation of T cell receptor and $\alpha 4\beta 1$ integrin can also lead to increased proliferation (Martsuyama et al 1989; Dang et al 1990; Nojima et al 1990; Shimizu et al 1990; Damle and Aruffo 1991; Yamada et al 1991). Indeed, the response depends on the state of activation of the cells, since resting T lymphocytes that have not yet encountered the antigen proliferate in response to the costimulus. On the other hand, antigen activated T cells may undergo apoptosis (Damle et al 1992a, 1992b and 1993). This phenomenon represents an important mechanism in immune response regulation.

$\alpha M\beta 2$ can also affect apoptosis in neuthophils as reported by Coxon et al (1996). Following phagocytosis of complement opsonized particles and release of reactive oxigen intermediates, neutrophils undergo apoptosis, a response important in downregulating acute inflammation and aiding in its resolution. The production of oxygen radicals following phagocytosis is responsible for the induction of neutrophil apoptosis. Phagocytosis, oxidative burst and subsequent apoptosis are impaired in αM deficient neutrophils indicating that this integrin, thanks to its ability to promote the oxidative burst can control apoptosis of these cells.

REFERENCES

Adams JC, Watt FM (1989) Fibronectin inhibits the terminal differentiation of human keratinocytes. Nature 340:307-309.

Agrez M, Chen A, Cone RI, Pytela R, Sheppard D (1994) The $\alpha V\beta 6$ integrin promotes proliferation of colon carcinoma cells through a unique region of the $\beta 6$ cytoplasmic domain. J Cell Biol 127:547-556.

Anwar ARF, Moqbel R, Walsh GM, Kay AB, Wardlaw AJ (1993) Adhesion to fibronectin prolongs eosinophil survival. J Exp Med 177:839-843.

Bates RC, Buret A, van Helden DF, Horton MA, Burns GF (1994) Apoptosis induced by inhibition of intercellular contact. J Cell Biol 125:403-415.

Bednarczyk JL, Kent Teague T, Wygant JN, Davis LS, Lipsky PE, McIntyre BW (1992) Regulation of T cell proliferation by anti-CD49d and anti-CD29 monoclonal antibodies. J Leu Biol 52:456-462.

Ben-Ze'ev A, Farmer S R, Penman S (1980) Protein synthesis requires cell surface contact while nuclear events respond to cell shape in anchorage dependent fibroblasts. Cell 21:365-372.

Benecke B J, Ben-Ze'ev A, Penman S (1978) The control of mRNA production, translation and turnover in suspended and re-attached anchorage dependent fibroblasts Cell 14:931-939.

Bohmer RM, Scharf E, Assoian RK (1996) Cytoskeletal integrity is required throughout the mitogen stimulation phase of the cell cycle and mediates the anchorage-dependent expression of cyclin D1. Mol Biol Cell 7:101-111.

Boldin MP, Goncharov TM, Goltsev YV, Wallach D (1996) Involvement of MACH, a novel MORT1/FADD-interacting protease, in Fas/APO-1- and TNF receptor-induced cell death. Cell 85:803-815.

Boudreau N, Sympson CJ, Werb Z, Bissel M (1995) Suppression of ICE and apoptosis in mammary epithelial cells by extracellular matrix. Science 267:891-893.

Bozzo C, Bellomo G, Silengo L, Tarone G, Silengo L, Altruda F (1997) Integrins and growth factors independently rescue human neuroblastoma cells from apoptosis induced by culture in suspension. Exp Cell Res (in press).

Brooks PC, Montgomery AM, Rosenfeld M, Reisfeld RA, Hu T, Klier G, Cheresh D (1994) Integrin αVβ3 antagonists promote tumor regression by inducing apoptosis of angiogenic blood vessels. Cell 79:1157-1164.

Bump NJ, Hackett M, Hugunin M, Seshagiri S, Brady K, Chen P, Ferenz C, Franklin S, Ghayur T, Li P et al (1995) Inhibition of ICE family proteases by baculovirus antiapoptotic protein p35. Science 269:1885-1888.

Cahill MA, Peter ME, Kischkel FC, Chinnaiyan AM, Dixit VM, Krammer PH, Nordheim A (1996) CD95 (APO-1/Fas) induces activation of SAP kinases downstream of ICE-like proteases. Oncogene 13:2087-2096.

Chen HC, Guan JL (1994) Association of focal adhesion kinase with its potential substrate phosphatidylinositol 3-kinase. Proc Natl Acad Sci USA 91:10148-52.

Chen Q, Kinch MS, Lin TS, Burridge K, Juliano RL (1994) Integrin-mediated cell adhesion activates mitogen-activated protein kinases. J Biol Chem 269:26602-26605.

Chong LD, Traynor-Kaplan A, Bokoch GM, Schwartz MA (1994) The small GTP-binding protein Rho regulates a phosphatidylinositol 4-phosphate 5-kinase in mammalian cells. Cell 79:507-513.

Clarke AS, Lotz MM, Chao C, Mercurio AM (1995) Activation of the p21 pathway of growth arrest and apoptosis by the beta 4 integrin cytoplasmic domain. J Biol Chem 270:22673-22676.

Coxon A, Rieu P, Barkalow FJ, Askari S, Sharpe AH, von Andrian UH, Arnaout MA, Mayadas TN (1996) A novel role for the β2 integrin CD11b/CD18 in neutrophil apoptosis: a homeostatic mechanism in inflammation. Immunity 5:653-666.

Crouch DH, Fincham VJ, Frame MC (1996) Targeted proteolysis of the focal adhesion kinase pp125 FAK during c-MYC-induced apoptosis is suppressed by integrin signaling. Oncogene 12:2689-2696.

Damle NK, Aruffo A (1991) Vascular cell adhesion molecule-1 induces T cell antigen receptor-dependent activation of human CD4+ T lymphocytes. Proc Natl Acad Sci USA 88:6403-6407.

Damle NK, Klussman K, Leytze G, Aruffo A, Linsley PS, Ledbetter JA (1993) Co-stimulation with integrin ligands intracellular adhesion molecule-1 or vascular cell adhesion molecule-1 augments activation-induced death of antigen-specific CD4+ T lymphocytes. J Immunol 151:2368-2379.

Damle NK, Klussman K, Linsley PS, Aruffo A (1992b) Differential co-stimulatory effects of adhesion molecules B7, ICAM-1, LFA-3, and VCAM-1 on resting and antigen-primed CD4+ T lymphocytes. J Immunol 148:1985-1990.

Damle NK, Klussman K, Linsley PS, Aruffo A, Ledbetter JA (1992a) Differential regulatory effects of ICAM-1 on co-stimulation by the CD28 counter-receptor B7. J Immunol 149:2541-2545.

Dang NH, Torimoto Y, Schlossman KSF, Morimoto C (1990) Human CD4 helper T cell activation: functional involvment of two distinct collagen receptors, 1F7 and VLA integrin family. J Exp Med 172:649-655.

Daub H, Weiss FU, Wallasch C, Ullrich A (1996) Role of transactivation of the EGF receptor in signalling by G-protein-coupled receptors. Nature 379:557-560.

Dike L E, Farmer S R (1988) Cell adhesion induces expression of growth-associated genes in suspension-arrested fibroblasts. Proc Natl Acad Sci USA 85:6792-6796.

El-Deiry WS, Harper JW, O'Connor PM, Velculescu VE, Canman CE, Jackman J, Pietenpol JA, Burrell M, Hill DE, Wang Y, et al (1994) WAF1/CIP1 is induced in p53-mediated G1 arrest and apoptosis. Cancer Res 54:1169-74.

Fang F, Orend G, Watanabe N, Hunter T, Ruoslahti E (1996) Dependence of cyclin E-Cdk2 kinase activity on cell anchorage. Science 271:499-502.

Farmer SR, Ben-Ze'av A, Benecke BJ, Penman S (1978) Altered translatability of messenger RNA from suspended anchorage-dependent fibroblasts: reversal upon cell attachment to a surface. Cell 15:627-37.

Folkman J, Moscona A (1978) Role of cell shape in growth control. Nature 273:345-349.

Fornaro M, Zheng DQ, Languino LR (1995) The novel structural motif Gln^{795}-Gln^{802} in the integrin β1C cytoplasmic domain regulates cell proliferation. J Biol Chem 270:24666-24669.

Frisch AM, Vouri K, Kelaita D, Sicks S (1996a) A role for Jun-N-terminal kinase in anoikis; suppression by Bcl-2 and crmA. J Cell Biol 135:1377-1382.

Frisch AM, Vouri K, Ruoslahti E, Chan-Yui PY (1996b) Control of adhesion-dependent cells survival by focal adhesion kinase. J Cell Biol 134:793-799.

Frisch SM, Hunter F (1994) Disruption of epithelial cell-matrix interactions induces apoptosis. J Cell Biol 124:619-626.

Giancotti FG, Ruoslahti E (1990) Elevated levels of the α5β1 fibronectin receptor suppress the transformed phenotype of CHO cells. Cell 60:849-859.

Groux H, Huet S, Valentin H, Pham D, Bernard A (1989) Suppressor effects and cyclic AMP accumulation by the CD29 molecule of CD4$^+$ lymphocytes. Nature 339:152-154.

Guadagno TM, Ohtsubo M, Roberts JM, Assoian RK (1993) A link between cyclin A expression and adhesion-dependent cell cycle progression. Science 262:1572-575.

Hannigan GE, Leung-Hagesteijn C, Fitz-Gibbon L, Coppolino MG, Radeva G, Filmus J, Bell JC, Dedhar S (1996) Regulation of cell adhesion and anchorage-dependent growth by a new beta1-integrin-linked protein kinase. Nature 379:91-6.

Hansen LK, Mooney DJ, Vacanti JP, Ingber DE (1994) Integrin binding and cell spreading on extracellular matrix act at different points in the cell cycle to promote hepatocyte growth. Mol Biol Cell 5:967-975.

Hermeking H, Eich D (1994) Mediation of c-Myc-induced apoptosis by p53. Science 265:2091-2093.

Howlett AR, Bailey N, Damsky C, Petersen OW, Bissell MJ (1995) Cellular growth and survival are mediated by β1 integrins in normal human breast epithelium but not in breast carcinoma. J Cell Sci 108:1945-1957.

Hungerford JE, Compton MT, Matter ML, Hoffstrom BG, Otey C (1996) Inhibition of p125Fak in cultured fibroblasts results in apoptosis. J Cell Biol 135:1383-1390.

Hurley RW, McCarthy JB, Verfaillie CM (1995) Direct adhesion to bone marrow stroma via fibronectin receptors inhibits hematopoietic progenitor proliferation. J Clin Invest 96:511-519.

Ingber DE (1990) Fibronectin controls capillary endothelial cell growth by modulating cell shape. Proc Natl Acad Sci USA 87:3579-3583.

Ingber DE, Prusty D, Frangioni JV, Cragoe EJ Jr, Lechene C, Schwartz MA (1990) Control of intracellular pH and growth by fibronectin in capillary endothelial cells. J Cell Biol 110:1803-1811.

Komiyama T, Ray CA, Pickup DJ, Howard AD, Thornberry NA, Peterson EP, Salvesen G (1994) Inhibition of interleukin-1 beta converting enzyme by the cowpox virus serpin CrmA. An example of cross-class inhibition. J Biol Chem 269:19331-19337.

Koopman G, Keehnen RMJ, Lindhout E, Newman W, Shimizu Y, van Seventer GA, de groot C, Pals ST (1994) Adhesion through the LFA (CD11a/CD18)-ICAM-1 (CD54) and VLA-4 (CD49d)-VCAM-1 (CD106) pathways prevents apoptosis of germinal center B cells. J Immunol 152:3760-3766.

Kubota Y, Kleinman HK, Martin GR, Lawley TJ (1988) Role of laminin and basement membrane in the morphological differentiation of human endothelial cells into capillary-like structures. J Cell Biol 107:1589-98.

Lees E (1995) Cyclin dependent kinase regulation. Curr Op Cell Biol 7:773-780.

Lowe S, Ruley HE (1993) Stabilization of the p53 tumor suppressor is induced by adenovirus-5 E1A and accompanies apoptosis. Genes Dev 7:535-545.

Mainiero F, Murgia C, Wary KK, Pepe A, Blumemberg M, Westwick JK, Der CJ, Giancotti FG (1997) The coupling of $\alpha6\beta4$ to Ras-MAPK pathways mediated by Shc controls keratinocyte proliferation. EMBO J 16:2365-2375.

Martsuyama T, Yamada A, Kay J, Yamada KM, Akiyama S, Schlossman SF, Morimoto C (1989) Activation of CD4 cells by fibronectin and anti-CD3 antibody: a synergistic effect mediated by VLA-5 fibronectin receptor complex. J Exp Med 170:1133-1148.

Matsuhisa T, Mori Y (1981) An anchorage dependent locus in the cell cycle for the growth of 3T3 cells. Exp Cell Res 135:393-398.

Matsumoto Y, Hiromatsu K, Sakai T, Kobayashi Y, Kimura Y, Usami J, Shinzato T, Maeda K, Yashikai Y (1994) Co-stimulation with LFA-1 triggers apoptosis in $\gamma\delta$ T cells on T cell receptor engagement. Eur J Immunol 24:2441-2445.

McNamee HM, Ingber DE, Schwartz MA (1993) Adhesion to fibronectin stimulates inositol lipid synthesis and enhances PDGF-induced inositol lipid breakdown. J Cell Biol 121:673-678.

Meredith JE Jr, Fazeli B, Schwartz MA (1993) The extracellular matrix as a cell survival factor. Mol Biol Cell 4:953-961.

Meredith JE Jr, Takada Y, Fornaro M, Languino LR, Schwartz MA (1995) Inhi-bition of cell cycle progression by the alternatively spliced integrin $\beta1C$. Science 269:1570-1572.

Miyamoto S, Teramoto H, Gutkind JS, Yamada KM (1996) Integrins can collaborate with growth factors for phosphorylation of receptor tyrosine kinases and MAP kinase activation: roles of integrin aggregation and occupancy of receptors. J Cell Biol 135:1633-1642.

Miyamoto S, Teramoto H, Gutkind JS, Yamada KM (1996) Integrins can collaborate with growth factors for phosphorylation of receptor tyrosine kinases and MAP kinase activation: roles of integrin aggregation and occupancy of receptors. J Cell Biol 135:1633-1642.

Miyashita T, Krajewski S, Krajewska M, Wang HG, Lin HK, Liebermann DA, Hoffman B, Reed JC (1994) Tumor suppressor p53 is a regulator of bcl-2 and bax gene expression in vitro and in vivo. Oncogene 9:1799-1805.

Miyashita T, Reed JC (1995) Tumor suppressor p53 is a direct transcriptional activator of the human bax gene. Cell 80:293-299.

Montgomery AM, Reisfeld RA, Cheresh D (1994) Integrin $\alpha V\beta3$ rescues melanoma cells from apoptosis in three-dimensional dermal collagen. Proc Natl Acad Sci USA 91:8856-8860.

Morgan DO (1995) Principles of Cdk regulation. Nature 374:131-134.

Mortarini R, Gismondi A, Maggioni A, Santoni A, Herlyn M, Anichini A (1995) Mitogenic activity of laminin on human melanoma and melanocytes: different signal requirements and role of beta 1 integrins. Cancer Res 55:4702-4710.

Mortarini R, Gismondi A, Santoni A, Parmiani G, Anichini A (1992) Role of the alpha 5 beta 1 integrin receptor in the proliferative response of quiescent human melanoma cells to fibronectin. Cancer Res 52:4499-4506.

Muzio M, Chinnaiyan AM, Kischkel FC, O'Rourke K, Shevchenko A, Ni J, Scaffidi C, Bretz JD, Zhang M, Gentz R, Mann M, Krammer PH, Peter ME, Dixit VM (1996) FLICE, a novel FADD-homologous ICE/CED-3-like protease, is recruited to the CD95 (Fas/APO-1) death-inducing signaling complex. Cell 85:817-827.

Nagata S, Golstein P (1995) The Fas death factor. Science 267:1449-1453.

Nojima Y, Humphries MJ, Mould AP, Komoriya A, Yamada KM, Schlossman SF, Morimoto C (1990) VLA-4 mediates CD3-dependent CD4+ T cell activation via the CS1 alternatively spliced domain of fibronectin. J Exp Med 172:1185-1192.

Oltvai ZN, Milliman CL, Korsmeyer SJ (1993) Bcl-2 heterodimerizes in vivo with a conserved homolog, Bax, that accelerates programmed cell death. Cell 74:609-619.

Otsuka H, Moskowitz MJ (1975) Arrest of 3T3 cells in G1 phase in suspension culture. J Cell Physiol 87:213-220.

Panayotou G, End P, Aumailley M, Timpl R, Engel J (1989) Domains of laminin with growth-factor activity. Cell 56:93-101.

Pasqualini R, Hemler M (1994) Contrasting roles for integrin β1 and β3 cytoplasmic domains in subcellular localization, cell proliferation and cell migration. J Cell Biol 125:447-460.

Patel T, Gores GJ, Kaufmann SH (1996) The role of proteases during apoptosis. FASEB J 10:587-597.

Raff MC (1992) Social controls on cell survival and cell death. Nature 356:397-400.

Rana B, Mischoulon D, Xie Y, Bucher NLR, Farmer S (1994) The extracellular matrix interactions can regulate the switch between growth and differentiation in rat hepatocytes: reciprocal expression of C/EBPa and immediate-early growth response transcription factors. Mol Cel Biol 14:5858-5869.

Re F, Zanetti A, Sironi M, Polentarutti N, Lanfrancone L, Dejana E, Colotta F (1994) Inhibition of anchorage-dependent cell spreading triggers apoptosis in cultured human endothelial cells. J Cell Biol 127:537-546.

Reed JC (1995) Regulation of apoptosis by bcl-2 family proteins and its role in cancer and chemoresistance. Curr Op Oncol 7:541-546.

Renshaw MW, Tokoz D, Schwartz MA (1996) Involvment of the small GTPases Rho in integrin-mediated activation of the mitogen-activated protein kinase. J Biol Chem 271:21691-21693.

Ruoslahti E Reed JC (1994) Anchorage dependence, integrins, and apoptosis. Cell 77:477-478.

Saelman EUM, Kely PJ, Santoro SA (1995) Loss of MDCK cell α2β1 integrin expression results in reduced cyst formation, failure of hepatocyte growth factor/scatter factor-induced branching morphogenesis, and increased apoptosis. J Cell Sci 108:3531-3540.

Sastry SK, Lakonishok M, Thomas DA, Muschler J, Horwitz AF (1996) Integrin alpha subunit ratios, cytoplasmic domains, and growth factor synergy regulate muscle proliferation and differentiation. J Cell Biol 133:169-184.

Schlaepfer DD, Hanks SK, Hunter T, van der Geer P (1994) Integrin-mediated signal transduction linked to Ras pathway by Grb2 binding to focal adhesion kinase. Nature 372:786-791.

Schreiner C, Fisher M, Hussein S, Juliano RL (1991) Increased tumorigenicity of fibronectin receptor deficient chinese hamster ovary cell variants. Cancer Res 51:1738-1740.

Schwartz MA (1993) Spreading of human endothelial cells on fibronectin or vitronectin triggers elevation of intracellular free calcium. J Cell Biol 120:1003-1010.

Schwartz MA, Lechene C, Ingber DE (1991) Insoluble fibronectin activates the Na/H antiporter by clustering and immobilizing integrin α5/β1, independent of cell shape. Proc Natl Acad Sci USA 88:7849-7853.

Shimizu Y, van Seventer GA, Horgan KJ, Shaw S (1990) Costimulation of proliferative responses of resting CD4+ T cells by the interaction of VLA-4 and VLA-5 with fibronectin or VLA-6 with laminin. J Immunol 145:59-67.

Shin S, Freedman VH, Risser R, Pollack R (1975) Tumorigenicity of virus-transformed cells in nude mice is correlated specifically with anchorage-independent growth in vitro. Proc Natl Acad Sci USA 72:4435-4439.

Stoker M, O'Neil C, Berryman S, Waxman V (1968) Anchorage and growth regulation in normal and virus transformed cells. Int J Cancer 3:683-693.

Sugahara H, Kanakura Y, Furitsu T, Ishihara K, Oritani K, Ikeda H, Kitayama H, Ishikawa J, Hashimoto K, Kanayama Y, Matsuzawa Y (1994) Induction of programmed cell death in human hematopoietic cell lines by fibronectin via its interaction with very late antigen 5. J Exp Med 179:1757-1766.

Sundberg C, Rubin K (1996) Stimulation of beta1 integrins on fibroblasts induces PDGF independent tyrosine phosphorylation of PDGF beta-receptors. J Cell Biol 132:741-752.

Symington BE (1990) Fibronectin receptor overexpression and loss of transformed phenotype in a stable variant of the K562 cell line. Cell Regul 1:637-648.

Symington BE (1992) Fibronectin receptor modulates cyclin-dependent kinase activity. J Biol Chem 267:25744-7.

Sympson CJ, Talhouk RS, Alexander CM, Chin JR, Clift SM, Bissell MJ, Werb Z (1994) Targeted expression of stromelysin-1 in mammary gland provides evidence for a role of proteinases in branching morphogenesis and the requirement for an intact basement membrane for tissue-specific gene expression. J Cell Biol 125:681-693.

Teague TK, McIntyre BW (1994) MAb 18D3 triggering of integrin beta 1 will prevent but not terminate proliferation of human T cells. Cell Adhes Commun 2:169-184.

Ticchioni M, Aussel C, Breittmayer JP, Maanie S, Pelassy C, Bernard A (1993) Suppressive effect of T cell proliferation via the CD29 molecule. J Immunol 151:119-127.

Varner JA, Emerson DA, Juliano RL (1995) Integrin $\alpha5\beta1$ expresion negatively regulates cell growth: reversal by attachment to fibronectin. Mol Biol Cell 6:725-740.

von der Mark K, Ocalan M (1989) Antagonistic effects of laminin and fibronectin on the expression of the myogenic phenotype. Differentiation 40:150-157.

Vuori K, Ruoslahti E (1993) Activation of protein kinase C precedes $\alpha5\beta1$ integrin-mediated cell spreading on fibronectin. J Biol Chem 268:21459-21462.

Vuori K, Ruoslahti E (1994) Association of insulin receptor substrate-1 with integrins. Science 266:1576-8.

Wang R, Kobayashi R, Bishop JM (1996) Cellular adherence elicits ligand-independent activation of the Met cell-surface receptor. Proc Natl Acad Sci USA 93:8425-8430.

Wary KK, Mainiero F, Isakoff SJ, Marcantonio E, Giancotti FG (1996) The adaptor protein Shc couples a class of integrins to the control of cell cycle progression. Cell 87:733-744.

Watt FM, Kubler MD, Hotchin NA, Nicholson LJ, Adams JC (1993) Regulation of keratinocyte terminal differentiation by integrin-extracellular matrix interactions. J Cell Sci 106:175-82.

White E (1996) Life, death and the pursuit of apoptosis. Genes Dev 10:1-15.

Xia Z, Dickens M, Raingeaud J, Davis RJ, Greenberg ME (1995) Opposing effects of ERK and JNK-p38 MAP kinases on apoptosis. Science 270:1326-31.

Xu LH, Owens LV, Sturge GC, Yang X, Liu ET, Craven RJ, Cance WG (1996) Attenuation of the expression of the focal adhesion kinase induces apoptosis in tumor cells. Cell Growth Differ 7:413-418.

Yamada A, Nojima Y, Sugita K, Dang NH, Schlossman SF, Morimoto C (1991) Crosslinking of VLA/CD29 molecule has a co-mitogenic effect with anti-CD3 on CD4 cell activation in serum-free culture system. Eur J Immunol 21:319-325.

Zhang Z, Vuori K, Reed JC, Ruoslahti E (1995) The $\alpha5\beta1$ integrin supports survival of cells on fibronectin and upregulates Bcl-2 expression. Proc Natl Acad Sci USA 92:6161-6165.

Zhu X, Assoian RK (1995) Integrin-dependent activation of MAPK: a link to shape-dependent cell proliferation. Mol Biol Cell 6:273-82.

Zhu X, Ohtsubo M, Bohmer RM, Roberts JM, Assoian RK (1996) Adhesion-dependent cell cycle progression linked to the expression of cyclin D1, activation of cyclin E-cdk2, and phosphorylation of the retinoblastoma protein. J Cell Biol 133:391-403.

CHAPTER 9

Integrin Regulation
of Hematopoietic Cell Function

C ell behavior and function results from the integration of multiple signaling pathways generated by cell surface receptors. Integrins transmit positional signals that are crucial in the organization of cells in tissues and in the control of their motility. These positional signals should integrate with signals from other receptors to regulate cell function in relation to the appropriate tissue context. Indeed it has become clear in the last few years that integrins establish dynamic and reciprocal interactions with other membrane receptors. Integrin activity can be radically modulated by stimulation of other receptors and they, in turn, can regulate different signaling pathways. A schema of the complex cross talk network between integrins and other membrane receptors is reported in Figure 9.1. The molecular basis of this crosstalk are still poorly defined and identification of the signaling molecules involved in these processes is a challenging theme for the ongoing research.

We have already discussed in chapters 7 and 8 some examples of how integrin signaling can affect hormone or growth factor receptor signaling in the expression of the differentiated phenotype and in the control of cell proliferation. In this chapter we will focus on the crosstalk between integrins and cell surface receptors in hemopoietic and immune cells. Cellular interactions in this system are rather well defined both at molecular and functional levels and several examples of how integrins can affect immune cell functions have been reported.

PLATELETS

Platelets are cells highly adapted for normal hemostasis and wound healing. They circulate as inactivated packets of vasoactive materials sequestered in granules dispersed within the platelet disc. With injury to a vessel, platelets are exposed to extracellular matrix components, namely fibrillar collagen, fibronectin and proteoglycans. This initiates the hemostatic cascade in which platelets undergo three important reactions: 1) adhesion and shape change, 2) granule secretion and 3) aggregation, collectively referred to as platelet activation. Central to these functions is cell adhesion which is mediated by multiple specialized receptors that by linking platelets to each other and to specific components of vessel wall, initiate both physiological hemostasis and pathological thrombosis. A key adhesion molecule involved in platelet hemostatic function is $\alpha IIb\beta 3$ integrin also known as GPIIb-IIIa. Absence of this

Signal Transduction by Integrins, by Paola Defilippi, Angela Gismondi, Angela Santoni and Guido Tarone. © 1997 Landes Bioscience.

Fig. 9.1. Crosstalk
and signaling co-
operation between
integrin and mem-
brane receptors.
Integrins and cell
surface receptors
can affect each
other in multiple
ways depending on
the specific cellular
system. Integrin-
matrix interaction
can regulate other
receptor function
as reported for EGF
receptor (Cybulsky
et al 1994) or re-
ceptor expression
(right curved ar-
row). At the same
time there are ex-
amples on the abil-
ity of different cyto-
kine receptors to
regulate integrin
expression and
state of activation
(left curved arrow).
In addition the
pathways triggered
by both class of re-

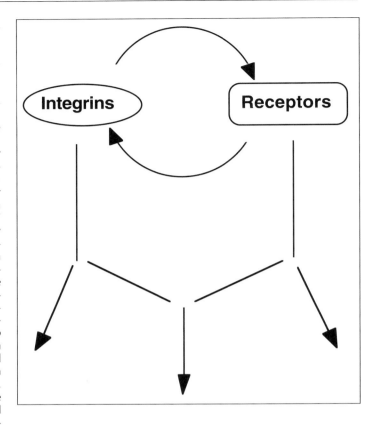

ceptors can intersecate (central arrow) or contribute to specific signaling events (lateral
arrows) necessary to generate a cellular response.

integrin in the genetic disease Glanzmann's
thrombasthenia causes defective platelet
adhesion and a bleeding disorder. In normal circulation and in resting platelets studied in vitro, this integrin is surface-exposed
but in a low-affinity state (Kd >> μM) with
respect to several soluble adhesive ligands
containing the sequence Arg-Gly-Asp such
as fibrinogen, von Willebrand factor,
vitronectin and fibronectin. However, when
platelets enter a vascular wound or encoun-
ter thrombin or other agonists, the receptor
is converted within seconds to a high-affinity state (Kd << μM) and ligand binding
ensues (Abrams et al 1994). A conformational change in both αIIb and β3 subunits
appears to be involved in the increased
integrin affinity (Sims et al 1991; Plow et al

1992). In addition, clustering of αIIbβ3
heterodimers has been observed upon
ligand binding (Isenberg et al 1987) and may
affect the strength and reversibility of the
adhesion process. The agonist-triggered biochemical events that directly modify the
conformation of αIIbβ3 are not known.
With the exception of collagen, which is
bound by several proteins present on the
platelet membrane, most agonists, including thrombin, epinephrine and ADP, act via
seven transmembrane-helix receptors
coupled to G proteins; moreover, direct activation of PKC by phorbol esters stimulates
fibrinogen binding and experiments with
inhibitors of various PTKs and PTPs, suggest that protein phosphorylation is a key
step in affinity modulation (Higashihara et
al 1992; Renduet al 1992; Shattil et al 1992).

NEUTROPHILS

Neutrophils play a central role in host defense against infections and are also major effector cells causing tissue damage in a variety of autoimmune and connective tissue disorders. During phagocytosis, many antimicrobial enzyme systems are activated and delivered into the phagocytic vesicles where they first kill and then digest the pathogens. These antimicrobial systems include an O_2-/H_2O_2-generating NADPH oxidase, myeloperoxidase, a variety of proteases and hydrolases. Neutrophils use several integrins for attachment to surfaces, such as endothelium or opsonized pathogens. During infection or inflammation, chemoattractants or cytokines bind to neutrophils and upregulate or "prime" their responsiveness. Priming can increase integrin receptor expression, but can also modulate receptor function resulting in altered integrin ligand-binding properties.

It has been well established that integrin receptors can regulate neutrophil function. Several findings demonstrate the cooperation between integrins and receptors for IgG (FcγR). There are three types of cell surface IgG receptors (FcγRI, FcγRII, FcγRIII) which bind IgG with different affinity and have different tissue distribution, but possess overlapping biological activities, such as phagocytosis, antibody dependent cellular cytotoxicity, mediator release and antigen presentation. Binding of IgG-opsonized particles to neutrophils does not require αMβ2 (CR3), but subsequent phagocytosis does (Fig. 9.2). Similarly, spreading of neutrophils to immune complex-coated particles or production of the potent neutrophil chemoattractant leukotriene B4 (LTB4) requires the participation of CR3. This LTB4 production is mediated by the interaction between the receptor for aggregated IgG, FcγRII and αMβ2 (Graham et al 1993). Coligation of the low-affinity receptor for IgG (FcγRIII) and αMβ2 on neutrophils activates a pertussis toxin-insensitive respiratory burst which is dependent on cytoskeleton integrity and activation of protein tyrosine kinases (Zhou et al 1994).

Several evidence indicates that FcγR may also interact with β3 integrins. Thus, the leukocyte response integrin (LRI), a unique β3 integrin that recognizes many Arg-Gly-Asp-containing extracellular matrix proteins through the association with the IAP /CD47 protein, is capable not only of activating neutrophil respiratory burst, adhesion and chemotaxis, but also leads to increased phagocytosis mediated by Fcγ receptors (Senior et al 1992; Gresham et al 1992).

In addition to the receptors for IgG, β2 integrins on neutrophils also cooperate with other surface receptors in the control of their functions. The best studied example is provided by the requirement of β2 integrins to make neutrophils responsive to TNFα and other inflammatory cytokines (Nathan 1987, 1989; Nathan et al 1989; Nathan et al 1989; Richter et al 1990). TNFα-treated neutrophils lower their cAMP (Nathan and Sanchez, 1990), reorganize their actin-based cytoskeleton to form focal adhesions (Nathan and Sanchez, 1990), spread out (Nathan and Sanchez, 1990), discharge proteases from their granules (Richter et al 1990) and mount a strong and prolonged respiratory burst (Laudanna et al 1990; Nathan, 1987; Nathan and Sanchez, 1990; Nathan et al 1989; Shappell et al 1990). Each of these responses depends on the expression of β2 integrins by neutrophils, as shown by the fact that they are blocked by specific antibodies against this family of integrins and are not seen in neutrophils from LAD patients, deficient for the expression of all leukocyte integrins that share the β2 chain. In neutrophils adherent to surface-coated with β2 ligands (i.e., fibrinogen) or anti-β2 antibodies, TNFα also initiates a series of signaling events such as oscillations of cytosolic free Ca^{2+} (Richter et al 1990), chloride ion efflux (Menegazzi et al 1996) and protein tyrosine phosphorylation (Fuortes et al 1993, 1994; Berton et al 1994). The generation of calcium signal enables TNFα to induce degranulation in adherent neutrophils (Richter et al 1990), whereas the Cl⁻ efflux induced by occupancy of the TNFα receptor p55 and engagement of β2 integrins

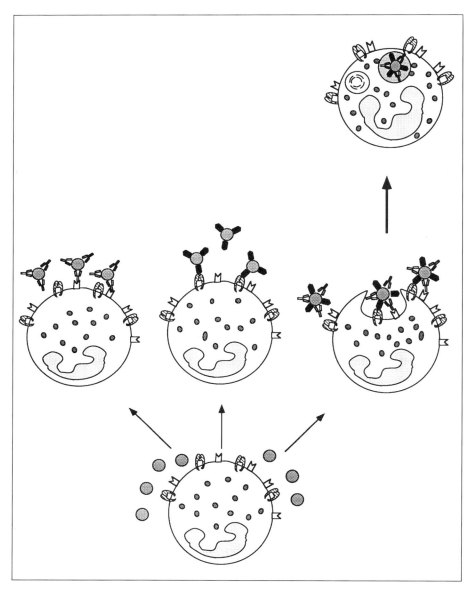

Fig. 9.2. Phagocytosis, but not binding of IgG-opsonized particles by neutrophils requires αμβ2 integrin. ⬆ and Σ represent C3bi and the FcγR, respectively.

is involved in the control of neutrophil adherence, spreading and activation of oxidative burst (Menegazzi et al 1996). TNFα-induced tyrosine phosphorylation also participates in the transduction of signals that direct the cells to spread on a biological surface and to undergo a respiratory burst, as shown by the ability of PTK inhibitors to block hydrogen peroxide production and spreading of neutrophils in response to TNFα (Fuortes et al 1993).

TNFα triggers a β2-integrin dependent activation of the Src family kinases, Fgr, Lyn and Hck, and tyrosine phosphorylation of paxillin; p125Fak activation, however, does not occur in these conditions (Fuortes et al 1994; Berton et al 1994 ; Graham et al 1994; Yan et al 1995; Yan et al 1996). The dependence of this response on β2 integrins has been shown by the failure of neutrophils from LAD patients to phosphorylate these substrates and by the ability of anti-β2 antibodies to abolish it. Paxillin tyrosine phosphorylation requires actin polymerization and occurs only in neutrophils that are both stimulated with TNFα and allowed to adhere. In adherent neutrophils, it is also induced by other stimuli including phorbol esters, the chemoattractant FLMP and opsonized bacteria (Fuortes et al 1994). In contrast, β2 integrins are sufficient to trigger protein tyrosine phosphorylation and activation of Fgr, Lyn, Hck and TNFα as well as other stimuli such as phorbol esters and FLMP enhance this response (Berton et al 1994; Yan et al 1995; Yan et al 1996). The functional role played by the Src family kinases Hck and Fgr in integrin signaling in neutrophils has been demonstrated by the use of knock-out mice (Lowell et al 1996). Neutrophils from double mutants Hck$^{-/-}$ Fgr$^{-/-}$ are defective in adhesion-dependent respiratory burst mediated by crosslinking of integrin receptors with either ECM proteins or surface-bound anti-integrin antibodies. This functional defect is the result of defective cell spreading. Single mutant Hck-/- or Fgr-/- neutrophils are completely normal. The mechanisms by which β2 integrins capacitate neutrophils to respond to

TNFα or other inflammatory stimuli are largely undefined. It is conceivable that the β2-dependent response of neutrophils results from the ability of TNFα or other agonists to generate a signal leading to integrin activation and increased affinity/avidity for their ligands; nonetheless, it cannot be excluded the possibility that it reflects a true costimulatory event which requires ligation of both β2 integrins and agonist receptors (Fig. 9.3).

Selectins, a family of proteins which participate in the process of leukocyte-endothelium attachment and their counter-receptors also crosstalk with β2 integrins. Adhesion of neutrophils to purified P and E selectins potentiates β2-dependent phagocytosis of unopsonized zymosan particles (Cooper et al 1994). Moreover, coligation of L-selectin and αMβ2 receptors stimulates neutrophil peroxide anion production (Crockett-Torabi et al 1995). These selectin-mediated costimulatory effects are associated with increased β2 integrin adhesiveness and the expression of the activation-dependent epitope detected by the mAb 24 on the β2 integrin chain (Cooper et al 1994; Simon et al 1995). Recently, similar to chemoattractants, L selectin was observed to impair the actin polymerizing capacity of β2 integrins (Ng-Sikorski et al 1996). This L selectin-induced effect was accompanied by a reduction in the β2-triggered tyrosine phosphorylation indicating that L selectin engagement can affect the β2 signal transduction pathway. This finding suggests that L selectin is responsible not only for the rolling that precedes the firm β2 integrin-mediated adhesion of neutrophils to the endothelium, but also for a negative feedback mechanism whereby firm adhesion is counteracted. This latter property of L selectin may be responsible for directing the neutrophil to a specific site.

Evidence is also available on the ability of β1 integrins to modulate neutrophil functions. Laminin promotes phagocytosis by primed neutrophils and primes neutrophil oxidative burst (Pike et al 1986). Furthermore, during alteration of oxygen tension,

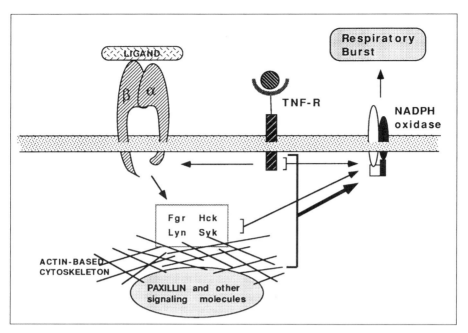

Fig. 9.3. Integrin-triggered intracellular events in neutrophils. Agonists such as TNFα generate a signal leading to integrin activation. Integrin clustering triggers activation of Src family (Fgr, Hck, Lyn) and Syk kinases. TNFα and/or tyrosine kinases can promote a partial assembly of NADPH oxidase (thin arrows). The full assembly of NADPH oxidase and the activation of respiratory burst requires the formation of an actin-based cytoskeleton (thick arrows).

a condition that often occurs at the sites of acute inflammation, fibronectin and laminin have been shown to control the expression on neutrophils of the αMβ2 integrin, the low-affinity receptor for IgG (FcγRIII/CD16) and the cytokine receptors TNFαR, IL-8R, IL-1R type I (Simms and D'Amico, 1995; 1996). The regulation of cytokine receptor expression by ECM proteins requires integrin signaling via α5β1 and α6β1 and in particular protein tyrosine kinase activation and intracellular calcium influx.

EOSINOPHILS

Eosinophils are predominantly tissue cells located in the skin or at the mucosal sites and play an important role in the pathogenesis of allergic diseases and in host resistance to parasitic infections. Their activation results in production of lipid mediators and cytokines and leads to release of granule proteins, such as major basic protein, eosinophil peroxidase, eosinophil cationic protein and eosinophil-derived neurotoxin. These basic proteins act as cytotoxins on a variety of cell types including pneumocytes, tracheal epithelium and parasites, thus contributing not only to defense against helminthic parasites, but also to tissue dysfunction damage in allergic diseases.

Eosinophils express several integrins mainly belonging to β1 and β2 families which have been described to affect their functions. Ligation of α4β1, αLβ2 and αMβ2 by antibodies triggers both eosinophil respiratory burst and spreading (Laudanna et al 1993) Similarly, α4β1-mediated adhesion to VCAM-1 activates a modest but significant amount of superoxide anion, which can be enhanced in response to the FLMP chemoattractant (Nagata et al 1995). Interestingly, VCAM-1-mediated stimulation of eosinophil O_2- generation was inhibited by the tyrosine kinase inhibitor genistein, suggesting a role for PTK activation in the enhancing effect. In addition, evidence has been reported that eosinophils adherent to fibronectin produce GM-CSF without exogenous stimuli (Anwar et al 1993) and release leukotriene C4 (LTC4), in response to PAF or calcium ionophore A23187 (Anwar et al 1994; Munoz et al 1996), and peroxidase, when stimulated with FLMP plus cytochalasin B (Neely et al 1994).

As shown for neutrophils, $\alpha M\beta2$-mediated cellular adhesion primes eosinophils for inflammatory mediator release in response to cytokines or chemoattractants; O_2- generation in response to TNFα and platelet-activating factor (PAF) has been observed when eosinophils are plated on fibrinogen, a ligand for $\alpha M\beta2$, but not with cells in suspension (Dri et al 1991); in addition, degranulation and peroxide production of stimulated eosinophils can be inhibited by blocking cellular adhesion with antibodies against $\alpha M\beta2$ (Horie and Kita, 1994). A crucial role for $\beta2$ integrins has been also reported in IgG-mediated activation of eosinophils. Thus, degranulation induced by immobilized IgG is inhibited by an antibody to $\beta2$ and is synergistically enhanced by fibrinogen, when it is coimmobilized with IgG to plates (Kaneko et al 1995). This cellular adhesion seems to affect early signaling events because anti-$\beta2$ antibodies abolished the production of inositol phosphates.

Evidence is available that integrins may also negatively regulate eosinophil functions. An early study has reported that in vitro superoxide production is inhibited when eosinophils are plated on monolayers of endothelial cells (Dri et al 1991). More recently, eosinophil adhesion to ECM proteins such as laminin or fibronectin, has been found to impair agonist (PAF, C5a)-induced granule exocytosis (Kita et al 1996). ECM-induced attenuation of degranulation is accompanied by reduced generation of inositol phosphates, strongly suggesting that

it is attributable to alterations in intracellular signaling rather than interference with the degranulation process itself.

BASOPHILS AND MAST CELLS

Basophils and mast cells play a central role both in inflammatory and immediate allergic reactions. The aggregation of IgE bound to high-affinity receptors (FcϵRI) on the surface of these cells causes degranulation and release of many potent inflammatory mediators including histamine, proteases, chemotactic factors, arachidonic acid metabolites and cytokines. Results indicate that integrins may be major regulators of mast cell and basophil proliferation and function. Integrin-mediated adherence of bone marrow mast cells to vitronectin augments IL-3-induced proliferation (Bianchine et al 1992). However, plating the cells on vitronectin is not sufficient to support cell growth; therefore, as for other cell types, a consequence of adherence is the elaboration of comitogenic signals that enhance IL-3-induced cell proliferation.

Adherence of basophils to fibronectin modulates the extent of cell degranulation. Binding of the rat basophilic leukemia cell line RBL-2H3 to immobilized fibronectin results in enhanced histamine release induced through FcϵRI stimulation or by the calcium ionophore A23187 (Hamawy et al 1992), which is associated with a dramatic change in cell shape, granule distribution and cytoskeleton organization. Integrins may also negatively regulate basophil granule exocytosis. Monoclonal antibodies to $\alpha M\beta2$, but not $\alpha L\beta2$, have been reported to inhibit IL-3-enhanced FcϵRI-mediated release of histamine (Watanabe et al 1992).

The mechanisms by which ECM proteins modulate basophil and mast cell functions remain to be elucidated (Fig. 9.4). Until recently, the effects of adherence were believed to be mainly related to changes in cell shape and cytoskeleton reorganization. Recent evidence, however, indicates that adhesion to fibronectin and FcϵRI aggregation synergistically regulate tyrosine phosphorylation of several proteins including

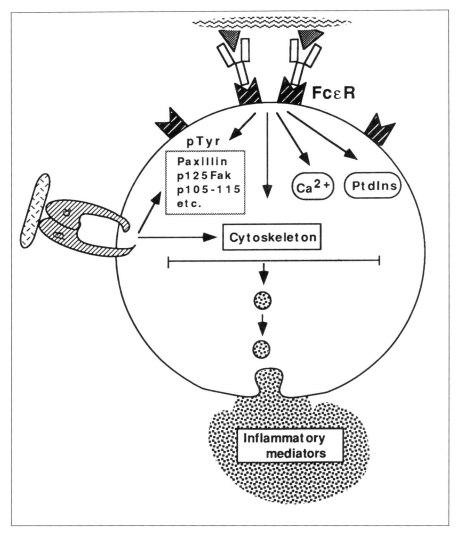

Fig. 9.4. Fibronectin enhances FcεRI-triggered basophil degranulation. Costimulation of degranulation is associated with cytoskeleton reorganization and synergistic modulation of tyrosine phosphorylation of several proteins including paxillin, p125Fak and p105-115.

paxillin, p125Fak and pp105-115 (Hamawy et al 1993a; 1993b). These data suggest that integrins regulate basophil and mast cell function by modulating intracellular signaling events.

MONOCYTES
AND MACROPHAGES

Cells from the monocyte-macrophage lineage play an essential role in the disposal of foreign agents or altered molecules, in the initiation and mediation of immune response and the inflammatory reaction; furthermore, they are involved in the repair process following various forms of tissue injury.

Several studies indicate that integrins by interaction with several ligands may modulate monocyte functional activities, such as phagocytosis, and synthesis of inflammatory mediators and cytokines. In monocytes, fibronectin stimulates CR1/CR3 and Fcγ receptor-mediated phagocytosis which may facilitate clearance of opsonized pathogens (Pommier et al 1983; Wright et al 1983; 1984); in addition, fibronectin is a strong inducer of GM-CSF (Thorens et al 1987) as well as of other cytokines (see chapter 7) and, in response to opsonized zymosan particles, increases the synthesis of PAF–a major lipid mediator implicated in the pathogenesis of many inflammatory reactions, including septic shock, anaphylaxis, asthma and immune-mediated tissue injury. These findings suggest that fibronectin, which accumulates in tissues during inflammation, may act as a proinflammatory mediator by stimulating monocyte-macrophage activity.

β2 integrins, namely αMβ2 and αXβ2, are the major receptors capable of delivering costimulatory signals in cells of monocyte-macrophage lineage. Antibody-mediated ligation of αMβ2 strongly potentiates LPS-induced expression of TNFα and tissue factor, an effector molecule involved in the initiation of coagulation cascade (Fan and Edgington, 1991; Fan and Edgington, 1993). In addition, as previously shown for fibronectin, αMβ2 acts in concert with a β-glucan receptor to induce PAF synthesis in response to opsonized zymosan particles (Elstad et al 1994). Recently, αMβ2 and αXβ2 have been reported to regulate monocyte activation through a novel interaction with the low-affinity receptor for IgE, CD23. This receptor, widely distributed among hematopoietic cells and implicated in the regulation of many immune and inflammatory responses, likely shares a close or identical epitope with the coagulation factor X, one of the ligands of αMβ2 and αXβ2. Thus, binding of recombinant CD23 to β2 integrins triggers monocytes to release proinflammatory mediators, such as nitric oxide, hydrogen peroxide and cytokines (IL-1β, IL-6 and TNFα) (Leocoanet-Henchoz et al 1995).

Integrins may also affect macrophage functions that limit inflammatory tissue injury and promote resolution of inflammation. Thus, the phagocytic clearance of se-

nescent neutrophils, a process which involves the recognition of neutrophils undergoing apoptosis, by the macrophage αvβ3 integrin (Savill et al 1990), is potentiated by thrombospondin via interaction with GPIV (CD36) (Savill et al 1992). This observation implies that proteins which are frequently incorporated into the ECM at inflamed sites may limit the toxic potential of neutrophils in inflammation by regulating the interaction between macrophages and apoptotic neutrophils.

T LYMPHOCYTES

T lymphocyte activation is achieved by the coordination of adhesion and signaling receptors expressed on the T cell surface binding to their counter-receptors expressed on the antigen-presenting cell or target cell. The specificity of T cells for antigen-bearing cells comes from recognition by the T cell receptor (TCR) of small peptides bound to major histocompatibility complex (MHC) molecules on antigen-presenting cells. However, contact between the TCR and the peptide-MHC complex is insufficient to trigger T lymphocyte activation and subsequent T lymphocyte effector functions. Coreceptors contribute to T lymphocyte response by increasing the adhesion between the T lymphocyte and the antigen-bearing cell, and by transmitting signals which augment the TCR-initiated activation cascade. A two-way functional interaction exists between TCR and coreceptors, in that TCR ligation results in enhanced binding between the coreceptor and its counter-receptor or ligand. This inside-out signaling has been well documented for many integrins expressed by T lymphocytes (Fig. 9.5). In fact, integrins on resting T cells exist in a low-avidity state and a high-avidity state is rapidly and transiently induced upon cell activation through the TCR/CD3 complex as well as a number of different surface receptors (Collins et al 1994). The mechanisms by which the cellular environment alters the functional state of integrins are still poorly understood. However, changes in receptor affinity involving integrin stabilization in an active conformation,

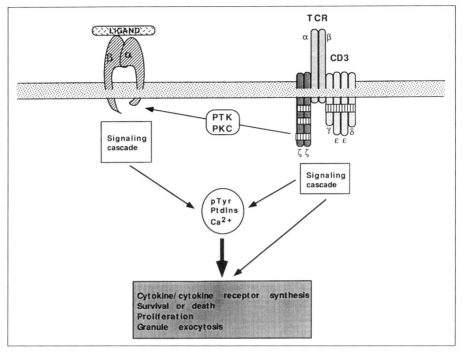

Fig. 9.5. Integrin-mediated modulation of T cell functions. TCR engagement generates intracellular biochemical events leading to integrin activation (inside-outside signaling). Integrin clustering initiates a signaling cascade which can affect TCR signaling in the expression of T cell functions.

association with cytoskeletal components or other molecules, and receptor clustering are likely to be involved (see chapter 1).

Much evidence is available for integrins as costimulatory receptors on T lymphocytes (Fig. 9.5). Early studies have indicated the important contribution of αLβ2 (LFA-1)/ICAM-1 interactions in antigen presentation. By the use of DNA-mediated gene transfer, in the absence of high levels of MHC expression or other accessory molecules on the antigen-presenting cells, ICAM-1 has been demonstrated to be a decisive factor in determining whether a T cell response occurs (Altmann et al 1989). In addition, ICAM-1 defective variants of B cells were found to have an impaired ability to present multiple antigens to T cells which could be restored by transfection with ICAM-1 cDNA (Dang et al 1990). Similarly,

a significant decrease has been observed in the antigen concentration required to activate T cells in the presence of immobilized ICAM-1 (Moy et al 1992). With naive T cells, ICAM-1 alone is not sufficient for optimal initiation of a T helper cell response, which is achieved by the coexpression on the antigen-presenting cells of the accessory molecule B7 (Dubey et al 1995). Interestingly, recent evidence indicates that the expression of B7 counter-receptors, CTL-4 and CD28, is upregulated upon costimulation of T lymphocytes with integrin ligands ICAM-1 or VCAM-1 (Damle et al 1994).

A large body of data is available on the ability of integrins to enhance TCR-mediated activation. Both ligand-induced or antibody-mediated cross linking of LFA-1, α3β1, α4β1, α5β1, α6β1 or α4β7 enhances TCR-mediated T cell proliferation (Carrera et al 1988; Matsuyama et al 1989; Wacholtz et al 1989; Nojima et al 1990; Shimizu et al 1990; Davis et al 1990; van Seventer et al 1990; Damle and Aruffo, 1991; van Seventer et al 1991; Burkly et al 1991; van Seventer et al 1992; Sarnacki et al 1992; Damle et al 1992; Hernandez-Caselles et al 1993; Ennis et al 1993). LFA-1 and $\alpha4\beta1$ integrins, not only potentiate TCR-induced growth, but also TCR-mediated activation-dependent death of T lymphocytes (Damle et al 1993), depending on the state of activation of T cells.

Integrins have been also shown to costimulate T cell effector functions such as cytokine production and cytotoxicity and to enhance the expression of several surface receptors including the α chain of IL-2R, the transferrin receptor CD71 and the early activation antigen CD69 (Yamada et al 1991; Van Seventer et al 1991; Sarnacki et al 1992; Hernandez-Caselles et al 1993). Thus, in the mouse, $\alpha v\beta3$ via interaction with fibronectin and vitronectin enhances IL-4 and IL-2 production by T cells expressing the γ/δ form of TCR and by a helper T cell hybridoma, respectively (Roberts et al 1991; Takahashi et al 1991). In human T cells, LFA-1, $\alpha4\beta1$ and $\alpha5\beta1$ can provide a costimulus for the release of multiple cytokines including IL-2, IL-4, TNFα, GM-CSF and IFNγ and the induction of the transcription factors NF-AT, AP-1 and NF-kB (Wacholtz et al 1989; Yamada et al 1991; Van seventer et al 1991; Semnani et al 1994; Udagawa et al 1996).

Several lines of evidence have been accumulated on the role of both $\beta1$ and $\beta2$ integrins in T cell-mediated cytotoxicity. The primary importance of $\alpha L\beta2$ (LFA-1)/ ICAM-1 adhesive interaction in the function of cytolytic T cells has been first demonstrated by observations that antibodies against LFA-1 and ICAM-1 block cytotoxic T cell-mediated killing (Makgoba et al 1988). More recently, it has been shown that purified ICAM-1 through interaction with LFA-1 facilitates the exocytosis of granules containing the cytotoxic mediators, by T cell clones, when coimmobilized with substimulatory amounts of anti-CD3 antibod-

ies. This enhanced response requires co-immobilization of ICAM-1 and anti-CD3 on the same surface suggesting that ICAM-1 transmits a very localized signal into the cell (Berg and Ostergaard, 1995). Synergism with TCR in the release of cytolytic granules by T lymphocytes has been also reported for fibronectin and other Arg-Gly-Asp-containing proteins, such as vitronectin and fibrinogen, both in mouse and human cells (Takahashi et al 1991; Ybarrondo et al 1994; Ostergaard and Ma 1995).

Among the signaling events that accompany the enhancement of T cell-mediated cytotoxic function, the stimulatory activity of fibronectin is associated with its ability to act in concert with the TCR/CD3 pathway to induce $Ins(1,4,5)$ P_3 generation (O'Rourke and Mescher, 1992), while ICAM-1 has been found to potentiate CD3-induced tyrosine phosphorylation and calcium flux (Berg and Ostergaard, 1995).

B LYMPHOCYTES

B lymphocytes are the only cells capable of producing antibodies. The interaction of antigen with membrane-bound immunoglobulin initiates the sequence of B cell activation, which culminates in the development of effector cells that actively secrete antibody molecules.

Few data are available on the ability of integrins to affect B cell proliferation, death and function, although several reports suggest a pivotal regulatory role of $\beta1$ integrins in human B cell ontogeny by mediating cell-cell and cell-stroma interaction of B cell precursors (Ryan and Tang, 1995). As shown for many integrins which deliver comitogenic signals on T cells, antibody-mediated ligation of $\alpha X\beta2$ on activated tonsillar B cells enhances PMA-induced proliferation and this response is reverted by fibrinogen (Postigo et al 1991). β_2 integrins, moreover, prevent apoptosis of germinal center B cells, and it has been suggested that this mechanism may play a relevant role in B cell selection (Koopman et al 1994).

Regarding $\beta1$ integrins, fibronectin via interaction with $\alpha4\beta1$ was found to be

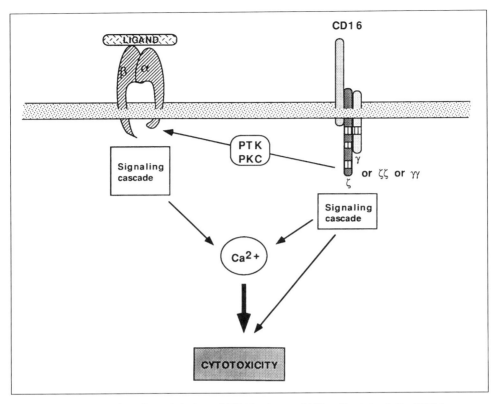

Fig. 9.6. Integrin-mediated modulation of CD16-triggered NK cell cytotoxicity. CD16 engagement generates intracellular biochemical events leading to integrin activation (inside-outside signaling). Integrin clustering initiates a signaling cascade which can affect CD16 signaling in the expression of NK cytotoxic function.

required together with IL-6, a major cytokine involved in plasma cell differentiation, to induce the terminal maturation of bone marrow-derived B cells capable of spontaneous and high rate immunoglobulin secretion (Roldan et al 1992). The mechanism by which this maturative event is triggered by fibronectin plus IL-6 is unknown. The possibility that they coordinately deliver a signal to B cells in a manner similar to that proposed for fibroblast growth regulation by basic FGF and ECM heparin-sulfate has been suggested. Moreover, as fibronectin produced by stromal cells mediates the inductive effect and stromal cells are also the source of IL-6, it is possible that the adhesive interaction plays a costimulatory role with respect to cytokine secretion and response.

NATURAL KILLER (NK) CELLS

NK cells are a small lymphocyte subpopulation endowed with the capacity of killing an array of target cells in a nonMHC-restricted or antibody-dependent manner. Since they can produce a variety of cytokines, they also display regulatory functions.

There are several reports suggesting the participation of β1 and β2 integrins in the control of NK cell functions. One of the first observations on the role of β2 integrins has been provided by Timonen et al (1988) who demonstrated that antibodies directed against αL, αM, αX or β2 strongly inhibit binding and cytotoxicity of human NK cells; similarly, enhancement of NK and CD16 (FcγRIII)-mediated cytotoxicity has been observed using a three-cell experimental system, comprising NK cells, target cells and ICAM-1 transfectants as a source of co-stimulation (Chong et al 1994). There is also evidence that a peptide from ICAM-2, another $\alpha L\beta 2$ ligand, strongly increases their cytotoxic activity against leukemic target cells (Nortamo et al 1993). In addition to the ability to regulate the cytotoxic function of NK cells, $\beta 2$ integrins also costimulate cytokine production. Thus, an antibody directed against αL integrin subunit enhances TNFα production triggered by CD16 ligation (Melero et al 1993). This antibody induces a transient increase in intracellular calcium suggesting that this event may be one of the signals accounting for the costimulation of TNFα production.

With respect to $\beta 1$ integrins, antibody- and ligand-mediated cross linking of $\alpha 4\beta 1$ and $\alpha 5\beta 1$ receptors enhances both spontaneous and CD16-mediated cytotoxic activity (Palmieri et al 1995). Enhancement of antibody-dependent activity by $\alpha 4\beta 1$ and $\alpha 5\beta 1$ correlates with the ability of $\beta 1$ integrins to potentiate CD16-triggered calcium flux (Fig. 9.6) (Palmieri et al 1996). The ability of fibronectin to enhance the cytotoxic function of NK cells may be particularly relevant in regulating their antitumoral and antiviral activity in the tissue micro-enviroment.

REFERENCES

Abrams CS, Ellison N, Budzynski AZ, Shattil SJ (1994) Direct detection of activated platelets and platelet-derived micro-particles in humans. Blood 75:128-138.

Altmann DM, Hogg N, Trowsdale J, Wilkinson D (1989) Cotransfection of ICAM-1 and HLA-DR reconstitutes human antigen-presenting cell function in mouse L cells. Nature 338:512-517.

Anwar AR, Moqbel R, Walsh JM, Kay AB, Wardlaw AJ (1993) Adhesion to fibronectin prolongs eosinophil survival. J Exp Med 177:839-843.

Anwar AR, Walsh JM, Cromwell O, Kay AB, Wardlaw AJ (1994) Adhesion to fibronectin primes eosinophils via $\alpha 4\beta 1$ (VLA-4). Immunology 82:222-228.

Berg NN, Ostergaard HL (1995) Characterization of intercellular adhesion molecule-1 (ICAM-1)-augmented degranulation by cytotoxic T cells. ICAM-1 and anti-CD3 must be co-localized for optimal adhesion and stimulation. J Immunol 155:1694-1702.

Berton G, Fumagalli L, Laudanna C, Sorio C (1994) $\beta 2$ integrin-dependent protein tyrosine phosphorylation and activation of the Fgr protein tyrosine kinase in human neutrophils. J Cell Biol 126:1111-1121.

Bianchine PJ, Burd PR, Metcalfe DD (1992) IL-3-dependent mast cells attach to plate-bound vitronectin. Demonstration of augmented proliferation in response to signals transduced via cell surface vitronectin receptors. J Immunol 149:3665-3671.

Burkly LC, Jakubowski A, Newman BM, Rosa MD, Chi-Rosso G, Lobb RR (1991) Signaling by vascular cell adhesion molecule-1 (VCAM-1) through VLA-4 promotes CD3-dependent T cell proliferation. Eur J Immunol 21:2871-2875.

Carrera AC, Rincon M, Sanchez-Madrid F, Lopez-Botet M, de Landazuri MO (1988) Triggering of co-mitogenic signals in T cell proliferation by anti-LFA-1 (CD18, CD11a), LFA-3, and CD7 monoclonal antibodies. J Immunol 141:1919-1924.

Chong AS-F, Boussy IA, Jiang XL, Lamas M, Graf LH (1994) CD54/ICAM-1 is a co-stimulator of NK cell-mediated cytotoxicity. Cell Immunol 157:92-105.

Collins TL, Kassner PD, Bierer BE, Burakoff SJ (1994) Adhesion receptors in lymphocyte activation. Curr Opin Immunol 6:385-393.

Cooper D, Butcher CM, Berndt MC, Vadas MA (1994) P-selectin interacts with a $\beta 2$-integrin to enhance phagocytosis. J Immunol 153:3199-3209.

Crockett-Torabi E, Sulenbarger B, Smith CW, Fantone JC (1995) Activation of human neutrophils through L-selectin and MAC-1 molecules. J Immunol 154:2291-2302.

Cybulsky AV, McTavish AJ, Cyr MD (1994) Extracellular matrix modulates epidermal growth factor receptor activation in rat glomerular epithelial cells. J Clin Invest 94:68-78.

Damle NK, Aruffo A (1991) Vascular cell adhesion molecule 1 induces T-cell antigen receptor-dependent activation of CD4+ T lymphocytes. Proc Natl Acad Sci USA 88:6403-6407.

Damle NK, Klussman K, Leytze G, Aruffo A, Linsley PS, Ledbetter JA (1993) Co-stimulation with integrin ligands intercellular adhesion molecule-1 or vascular cell adhesion molecule-1 augments activation-induced death of antigen-specific CD4+ T lymphocytes. J Immunol 151:2368-2379.

Damle NK, Klussman K, Leytze G, Myrdal S, Aruffo A, Ledbetter JA, Linsley PS (1994) Costimulation of T lymphocytes with integrin ligands intercellular adhesion molecule-1 or vascular cell adhesion molecule-1 induces functional expression of CTLA-4, a second receptor for B7. J Immunol 152:2686-2697.

Damle NK, Klussman K, Linsley PS, Aruffo A (1992) Differential costimulatory effects of adhesion molecules B7, ICAM-1 LFA-3, and VCAM-1 on resting and antigen-primed CD4+ T lymphocytes. J Immunol 148:1985-1992.

Dang LH, Michalek MT, Takei F, Benaceraff B, Rock KL (1990) Role of ICAM-1 in antigen presentation demonstrated by ICAM-1 defective mutants. J Immunol 144:4082-4091.

Davis LS, Oppenheimer-Marks N, Bednarczyk JL, McIntyre BW, Lipsky PE (1990) Fibronectin promotes proliferation of naive and memory T cells by signaling through both the VLA-4 and VLA-5 integrin molecules. J Immunol 145:785-793.

Dri P, Cramer R, Spessotto P, Romano M, Patriarca P (1991) Eosinophil activation on biologic surfaces. Production of O2- in response to physiologic soluble stimuli is differentially modulated by extracellular matrix components and endothelial cells. J Immunol 147:613-620.

Dubey C, Croft M, Swain SL (1995) Co-stimulatory requirements of naive CD4+ T cells ICAM-1 or B7-1 can costimulate naive CD4 T cell activation but both are required for optimum response. J Immunol 155:45-57.

Elstad MR, Parker CJ, Cowley FS, Wilcox LA, McIntyre TM, Prescott SM, Zimmerman GA (1994) CD11b/CD18 integrin and β-glucan receptor act in concert to induce the synthesis of platelet-activating factor by monocytes. J Immunol 152:220-230.

Ennis E, Isberg RR, Shimizu Y (1993) Very late antigen 4-dependent adhesion and costimulation of resting human T cells by the bacterial β_1 integrin ligand invasin. J Exp Med 177:207-212.

Fan S-T, Edgington TS (1991) Coupling of the adhesive receptor CD11b/CD18 to functional enhancement of effector macrophage tissue factor response. J Clin Invest 87:50-57.

Fan S-T, Edgington TS (1993) Integrin regulation of leukocyte inflammatory functions. CD11b/CD18 enhancement of the tumor necrosis factor-a responses of monocytes. J Immunol 150:2972-2980.

Fuortes M, Jin W, Nathan C (1993) Adhesion-dependent protein tyrosine phosphorylation in neutrophils treated with tumor necrosis factor. J Cell Biol 120:777-784.

Fuortes M, Jin W, Nathan C (1994) β_2 integrin-dependent tyrosine phosphorylation of paxillin in human neutrophils treated with tumor necrosis factor. J Cell Biol 127:1477-1483.

Graham IL, Anderson DC, Holers VM, Brown EJ (1994) Complement receptor 3 (CR3, Mac-1, integrin $\alpha\mu\beta_2$, CD11b/CD18) is required for tyrosine phosphorylation of paxillin in adherent and nonadherent neutrophils. J Cell Biol 127:1139-1147.

Graham IL, Lefkowith JB, Anderson DC, Brown EJ (1993) Immune complex-stimulated neutrophil LTB4 production is dependent on β_2 integrins. J Cell Biol 120:1509-1517.

Gresham HD, Adams SP, Brown EJ (1992) Ligand binding specificity of the leukocytes response integrin expressed by human neutrophils. J Biol Chem 267:13895-13992.

Hamawy MM, Mergenhagen SE, Siraganian RP (1993) Cell adherence to fibronectin and the aggregation of the high-affinity immunoglobulin E receptor synergistically regulate tyrosine phosphorylation of 105-115-kDa proteins. J Biol Chem 268:5227-5233.

Hamawy MM, Mergenhagen SE, Siraganian RP (1993) Tyrosine phosphorylation of pp125FAK by the aggregation of high-affinity immunoglobulin E receptors requires cell adherence. J Biol Chem 268:6851-6854.

Hamawy MM, Oliver C, Mergenhagen SE, Siraganian RP (1992) Adherence of rat basophilic leukemia (RBL-2H3) cells to fibronectin-coated surfaces enhances secretion. J Immunol 149:615-621.

Hernandez-Caselles T, Rubio G, Campanero MR, del Pozo MA, Muro M, Sanchez-Madrid F, Aparicio P (1993) ICAM-3, the third LFA-1 counterreceptor, is a co-stimulatory molecule for both resting and activated T lymphocytes. Eur J Immunol 23:2799-2806.

Higashihara M, Takahata K, Kurokawa K, Ikebe M (1992) The inhibitory effect of okadaic acid on platelet function. FEBS Lett 307:206-210.

Horie S, Kita H (1994) CD11b/CD18 (Mac-1) is required for degranulation of human eosinophils by human recombinant granulocyte-macrophage colony-stimulating factor and platelet-activating factor. J Immunol 152:5457-5467.

Isenberg WM, McEver RP, Phillips DR, Shuman MA, Bainton DF (1987) The platelet fibrinogen receptor: an immunogold surface replica study of agonist-induced ligand binding and receptor clustering. J Cell Biol 104:1655-1663.

Kaneko M, Horie S, Kato M, Gleich GJ, Kita H (1995) A crucial role for β_2 integrin in the activation of eosinophils stimulated by IgG. J Immunol 155:2631-2641.

Kita H, Horie S, Gleich GJ (1996) Extracellular matrix proteins attenuate activation and degranulation of stimulated eosinophils. J Immunol 156:1174-1181.

Koopman G, Keehnen RRJ, Lindhout E, Newman W, Shimizu Y, van Seventer GA, de Groot C, Pals ST (1994) Adhesion through the LFA-1 (CD11a/CD18)-ICAM-1 (CD54) and the VLA-4 (CD49d)-VCAM-1 (CD106) pathways prevents apoptosis of germinal center B cells. J Immunol 152:3760-3767.

Laudanna C, Melotti P, Bonizzato C, Piacentini G, Boner A, Serra MC, Berton G (1993) Ligation of members of the β_1 or the β_2 subfamilies of integrins by anti-bodies triggers eosinophil respiratory burst and spreading. Immunology 80:273-280

Laudanna C, Miron S, Berton G, Rossi F (1990) Tumor necrosis factor-alpha/cachectin activate the O2- generating system of human neutrophils independently of hydrolysis of phosphoinositides and the release of arachidonic acid. Biochem Biophys Res Commun 166:308-315.

Lecoanet-Henchoz S, Gauchat J-F, Aubry J-P, Graber P, Life P, Paul-Eugene N, Ferrua B, Corbi AL, Dugas B, Plater-Zyberk C, Bonnefoy J-Y (1995) CD23 regulates monocyte activation through a novel interaction with the adhesion molecules CD11b-CD18 and CD11c-CD18. Immunity 3:119-125.

Li R, Nortamo P, Kantor C, Kovanen P, Timonen T, Gahmberg CG (1993) A leukocyte integrin binding peptide from intercellular adhesion molecule-2 stimulates T cell adhesion and natural killer cell activity. J Biol Chem 268:21474-21477.

Lowell CA, Fumagalli L, Berton G (1996) Deficiency of Src family kinases p59/61hck and p58c-fgr results in defective adhesion-dependent neutrophil functions. J Cell Biol 133:895-910.

Makgoba MW, Sanders ME, Luce GEG, Gugel EA, Dustin ML, Springer TA, Shaw S (1988) Functional evidence that intercellular adhesion molecule-1 (ICAM-1) is a ligand for LFA-1 in cytotoxic T cell recognition. Eur J Immunol 18:637-640.

Matsuyama T, Yamada A, Kay J, Yamada KM, Akiyama SK, Schlossman SF, Morimoto C (1989) Activation of CD4 cells by fibronectin and anti-CD3 antibody. J Exp Med 170:1133-1148.

Melero I, Balboa M, Alonso JL, Yague E, Pivel JP, Sanchez-Madrid F, Lopez-Botet M (1993) Signaling through the LFA-1 leukocyte integrin actively regulates intercellular adhesion and tumor necrosis factor-a production in natural killer cells. Eur J Immunol 23:1859-1865.

Menegazzi R, Busetto S, Dri P, Cramer R, Patriarca P (1996) Chloride ion efflux regulates adherence, spreading and respiratory burst of neutrophils stimulated by tumor necrosis factor-α (TNF) on biologic surfaces. J Cell Biol 135:511-522.

Moy VT, Brian AA (1992) Signaling by lymphocyte function-associated antigen 1 (LFA-1) in B cells: enhanced antigen presentation after stimulation through LFA-1. J Exp Med 175:1-7.

Munoz NM, Rabe KF, Neeley SP, Herrnreiter A, Zhu X, McAllister K, Mayer D, Magnussen H, Galens S, Leff AR (1996) Eosinophil VLA-4 binding to fibronectin augments bronchial narrowing through 5-lipoxygenase activation. Am J Physiol 270:587-594.

Nagata M, Sedgwick JB, Bates ME, Kita H, Busse WW (1995) Eosinophil adhesion to vascular cell adhesion molecule-1 activates superoxide anion generation. J Immunol 155:2194-2202.

Nathan C, Sanchez E (1990) Tumor necrosis factor and CD11/CD18 (β_2) integrins act synergistically to lower cAMP in human neutophils. J Cell Biol 111:2171-2181.

Nathan C, Srimal S, Farber C, Sanchez E, Kabbash L, Asch A, Gailit J, Wright SD (1989) Cytokine-induced respiratory burst of human neutrophils: dependence on extracellular matrix proteins and CD11/CD18 integrins. J Cell Biol 109:1341-1349.

Nathan CF (1987) Neutrophil activation on biological surfaces. Massive secretion of hydrogen peroxide in response to products of macrophages and lymphocytes. J Clin Invest 80:1550-1560.

Nathan CF (1989) Respiratory burst in adherent human neutrophils: triggering by colony stimulating factors CSF-GM and CSF-G. Blood 73:301-306.

Neeley SP, Hamann KJ, Dowling TL, McAllister KT, White SR, Leff AR (1994) Augmentation of stimulated eosinophil degranulation by VLA-4 (CD49d)-mediated adhesion to fibronectin. Am J Respir Cell Mol Biol 11:206-213.

Ng-Sikorski J, Linden L, Eierman D, Franzen L, Molony L, Andersson T (1996) Engagement of L-selectin impairs the actin polymerizing capacity of β_2-integrins on neutrophils. J Cell SCi 109:2361-2369.

Nojima Y, Humphries MJ, Mould AP, Komoriya A, Yamada KM, Schlossman SF, Morimoto C (1990) VLA-4 mediates CD3-dependent CD4+ T cell activation via the CS1 alternatively spliced domain of fibronectin. J Exp Med 172:1185-1192.

O'Rourke AM, Mescher MF (1992) Cytotoxic T-lymphocyte activation involves a cascade of signalling and adhesion events. Nature 358:253-255.

Ostergaard HL, Ma EA (1995) Fibronectin induces phosphorylation of a 120-kDa protein and synergizes with the T cell receptor to activate cytotoxic T cell clones. Eur J Immunol 25:252-256.

Palmieri G, Gismondi A, Galandrini R, Milella M, Serra A, De Maria R, Santoni A. The interaction of NK cells with extracellular matrix induces early intracellular signalling events and enhances cytotoxic functions. Nat Immun 15:47-153.

Palmieri G, Serra A, De Maria R, Gismondi A, Milella M, Piccoli M, Frati L, Santoni A (1995) Crosslinking of $\alpha_4\beta_1$ and $\alpha_5\beta_1$ fibronectin receptors enhances natural killer cell cytotoxic activity. J Immunol 155:5314-5322.

Pike MC, Wicha MS, Yoon P, Mayo L, Boxer LA (1989) Laminin promotes the oxidative burst in human neutrophils via increased chemoattractant receptor expression. J Immunol 142:2004-2011.

Plow EF, D'Souza SE, Ginsberg MH (1992) Ligand binding to GP IIb-IIIa: a status report. Semin Thromb Hemost 18:324-332

Pommier CG, Shinichi I, Fries LF, Takahashi T, Frank MM, Brown EJ (1983) Plasma fibronectin enhances phagocytosis of opsonized particles by human peripheral blood monocytes. J Exp Med 157:1844-1854.

Postigo AA, Corbi AL, Sanchez-Madrid F, de Landazuri MO (1991) Regulated expression and function of CD11c/CD18 integrin on human B lymphocytes. Relation between attachment to fibrinogen and triggering of proliferation through CD11c/CD18. J Exp Med 174:1313-1322.

Rendu F, Eldor A, Grelac F, Bachelot C, Gazit A, Gilon C, Levy-Toledano S, Levitzki A (1992) Inhibition of platelet activation by tyrosine kinase inhibitors. Biochem Pharmacol 44:881-888.

Richter J, Ng-Sikorski J, Olsson I, Andersson T (1990) Tumor necrosis factor-induced degranulation in adherent human neutrophils is dependent on CD11b/CD18-integrin-triggered oscillations of cytosolic free Ca2+. Proc Natl Acad Sci USA 87:9472-9476.

Roberts K, Yokoyama WM, Kehn PJ, Shevach EM (1991) The vitronectin receptor serves as an accessory molecule for the activation of a subset of γ/δ T cells. J Exp Med 173:231-240.

Roldan E, Garcia-Pardo A, Brieva JA (1992) VLA-4-fibronectin interaction is required for the terminal differentiation of human bone marrow cells capable of spontaneous and high rate immunoglobulin secretion. J Exp Med 175:1739-1747.

Ryan DH, Tang J (1995) Regulation of human B cell lymphopoiesis by adhesion molecules and cytokines. Leuk Lymphoma 17:375-389.

Sarnacki S, Begue B, Buc H, Le Deist F, Cerf-Bensussan N (1992) Enhancement of CD3-induced activation of human intestinal intraepithelial lymphocytes by stimulation of the β_7-containing integrin defined by HML-1 monoclonal antibody. Eur J Immunol 22:2887-2892.

Savill J, Dransfiel I, Hogg N, Haslett C (1990) Vitronectin receptor mediated phagocytosis of cells undergoing apoptosis. Nature 3:170-173.

Savill J, Hogg N, Ren Y, Haslett C (1992) Trombospondin cooperates with CD36 and the vitronectin receptor in macrophage recognition of neutrophils undergoing apoptosis. J Clin Invest 90:1513-1522.

Semnani RT, Nutman TB, Hochman P, Shaw S, van Seventer GA (1994) Costimulation by purified intercellular adhesion molecule 1 and lymphocyte function-associated antigen 3 induces distinct proliferation, cytokine and cell surface antigen profiles in human "naive" and "memory" CD4+ T cells. J Exp Med 180:2125-2135.

Senior RM, Gresham HD, Griffin GL, Brown EJ, Chung AE (1992) Entactin stimulates neutrophil adhesion and chemotaxis through interactions between its Arg-Gly-Asp (RGD) domain and the leukocyte response integrin. J Clin Invest 90:2251-2257.

Shappell SB, Toman C, Anderson DC, Taylor AA, Entman ML, Smith CW (1990) Mac-1 (CD11b/CD18) mediates adherence-dependent hydrogen peroxide production by human and canine neutrophils. J Immunol 144:2702-2711.

Shattil SJ, Cunningham M, Wiedmer T, ZhaoJ, Sims PJ, Brass LF (1992) Regulation of glycoprotein IIb-IIIa receptor function studied with platelets permeabilized by the pore-forming complement proteins C5b-9. J Biol Chem 267:18424-18431.

Shimizu Y, Van Seventer GA, Horgan KJ, Shaw S (1990) Costimulation of proliferative responses of resting CD4+ T cells by the interaction of VLA-4 and VLA-5 with fibronectin or VLA-6 with laminin. J Immunol 145:59-67.

Simms H, D'Amico R (1995) Regulation of polymorphonuclear neutrophil CD16 and CD11b/CD18 expression by matrix proteins during hypoxia is VLA-5, VLA-6 dependent. J Immunol 155:4979-4990.

Simms H, D'Amico R (1996) Regulation of polymorphonuclear leukocyte cytokine receptor expression. The role of altered oxygen tension and matrix proteins. J Immunol 157:3605-3616.

Simon SI, Burns AR, Taylor AD, Gopalan PK, Lynam EB, Sklar LA, Smith CW (1995) L-selectin (CD62L) crosslinking signals neutrophil adhesive functions via Mac-1 (CD11b/CD18) β_2-integrin. J Immunol 155:1502-1514.

Sims PJ, Ginsberg MH, Plow EF, Shattil SJ (1991) Effect of platelet activation on the conformation of the plasma membrane glycoprotein IIb-IIIa complex. J Biol Chem 266:7345-7352.

Takahashi K, Nakamura T, Adachi H, Yagita H, Okumura K (1991) Antigen-independent T cell activation mediated by very late activation antigen-like extracellular matrix receptor. Eur J Immunol 21:1559-1562.

Thorens B, Mermod J-J, Vassalli P (1987) Phagocytosis and inflammatory stimuli induce GM-CSF mRNA in macrophages through posttrascriptional regulation. Cell 48:671-679.

Timonen T, Patarroyo M, Gahmberg CG (1988) CD11a-c/CD18 and GP84 (LB-2) adhesion molecules on human large granular lymphocytes and their participation in natural killing. J Immunol 141:1041-1046.

Udagawa T, Woodside DG, McIntyre BW (1996) $\alpha_4\beta_1$ (CD49d/CD29) integrin costimulation of human T cells enhances transcription factor and cytokine induction in the absence of altered sensitivity to anti-

CD3 stimulation. J Immunol 157:1965-1972.

Van Seventer GA, Bonvini E, Yamada H, Conti A, Stringfellow S, June CH, Shaw S (1992) Costimulation of T cell receptor/ CD3-mediated activation of resting human CD4+ T cells by leukocyte function-associated antigen-1 ligand intercellular cell adhesion molecule-1 involves prolonged inositol phospholipid hydrolysis and sustained increase of intracellular Ca2+ levels. J Immunol 149:3872-3880.

Van Seventer GA, Newman W, Shimizu Y, Nutman TB, Tanaka Y, Horgan KJ, Gopal TV, Ennis E, O'Sullivan D, Grey H, Shaw S (1991) Analysis of T cell stimulation by superantigen plus major histocompatibility complex class II molecules or by CD3 monoclonal antibody: costimulation by purified adhesion ligands VCAM-1, ICAM-1, but not ELAM-1. J Exp Med 174:901-913.

Van Seventer GA, Shimizu Y, Horgan KJ, Shaw S (1990) The LFA-1 ligand ICAM-1 provides an important costimulatory signal for T cell receptor-mediated activation of resting T cells. J Immunol 144:4579-4586.

Wacholtz MC, Patel SS, Lipsky PE (1989) Leukocyte function-associated antigen 1 is an activation molecule for human T cells. J Exp Med 170:431-448.

Watanabe A, Tominaga T, Tsuji J, Yanagihara Y, Koda A (1992) Effects of anti-CD11a, anti-CD11b and anti-CD18 on histamine release from human basophils primed with IL-3. Int Arch Allergy Immunol 98:308-310.

Wright SD, Craigmyle LS, Silverstein SC (1983) Fibronectin and serum amyloid P component stimulate C3b- and C3bi- me-

diated phagocytosis in cultured human monocytes. J Exp Med 158:1338-1343.

Wright SD, Licht MR, Craigmyle LS, Silverstein SC (1984) Communication between receptors for different ligands on a single cell: ligation of fibronectin receptors induces a reversible alteration in the function of complement receptors on cultured human monocytes. J Cell Biol 99:336-339.

Yamada A, Nikaido T, Nojima Y, Schlossman SF, Morimoto C (1991) Activation of human CD4 T lymphocytes. Interaction of fibronectin with VLA-5 receptor on CD4 cells induces the AP-1 transcription factor. J Immunol 146:53-56.

Yan SR, Berton G (1996) Regulation of Src family tyrosine kinase activities in adherent human neutrophils. J Biol Chem 271:23464-23471.

Yan SR, Fumagalli L, Dusi S, Berton G (1995) Tumor necrosis factor triggers redistribution to a Triton X-100-insoluble, cytoskeletal fraction of β_2 integrins, NADPH oxidase components, tyrosine phosphorylated proteins, and the protein tyrosine kinase p58fgr in human neutrophils adherent to fibrinogen. J Leukoc Biol 58:595-606.

Ybarrondo B, O'Rourke AM, Brian AA, Mescher MF (1994) Contribution of lymphocyte function-associated-1/intercellular adhesion molecule-1 binding to the adhesion/signaling cascade of cytotoxic T lymphocyte activation. J Exp Med 179:359-363.

Zhuo M, Brown EJ (1994) CR3 (Mac-1, $\alpha\mu\beta_2$, CD11b/CD18) and FcγRIII cooperate in generation of a neutrophil respiratory burst: requirement for FcγRII and tyrosine phosphorylation. J Cell Biol 125:1407-1416.

INDEX

DEMCO